Computational Mechanics of the Classical Guitar

Rolf Bader

Computational Mechanics of the Classical Guitar

With 65 Figures

 Springer

Dr. Rolf Bader
University of Hamburg
Musikwissenschaftliches Institut
Neue Rabenstr. 13
20354 Hamburg, Germany
e-mail: R_Bader@t-online.de

Library of Congress Control Number: 2005925685

ISBN-10   3-540-25136-7 Springer Berlin Heidelberg New York
ISBN-13   978-3-540-25136-1 Springer Berlin Heidelberg New York

Springer is a part of Springer Science+Business Media
springer.com
© Springer-Verlag Berlin Heidelberg 2005
Printed in The Netherlands

Typesetting: by the author and TechBooks using a Springer LATEX macro package

Cover design: *design & production* GmbH, Heidelberg

Printed on acid-free paper   SPIN: 11399407   89/TechBooks   5 4 3 2 1 0

To my parents for their love and support in good and in bad times.
To Prof. Albrecht Schneider for his critical and broad ranged teachings about most I know about music, his support and his endless and patient flow of ideas, energy and wisdom.

Preface

Acknowledgements

Thanks to Dr. Maria Haase, Gunter Faust and the people at ISD at the University of Stuttgart for their teachings, their support and encouragement.

Thank also to Elizabeth Hicks for most of the translation from German to English.

Contents

Glossary of Variables

Coordinates

Ω	domain
Γ	boundary
R_Γ	boundary condition
x	x-coordinate
y	y-coordinate
z	z-coordinate
u	displacement vector $\{r, s, t\}$
r or u_x	displacement in x-direction
s or u_y	displacement in y-direction
t or u_z	displacement in z-direction
n	unit outward normal vector

Matrices

D	n × m matrix of spatial vectors of the central points of the top plate elements
B	n × m matrix of spatial vectors of the central points of the back plate elements
Z	n × m matrix of spatial vectors of the central points of the rib elements
H	n × m matrix of spatial vectors of the central points of the neck elements
L	n × m × o matrix of spatial vectors of the central points of the enclosed air space elements
$\mathbf{S}_s$	n × s matrix of string elements of a string s

$\mathbf{D_u}$ n × m matrix of displacements of the central points
of the top plate elements

$\mathbf{B_u}$ n × m matrix of displacements of the central points
of the back plate elements

$\mathbf{Z_u}$ n × m matrix of displacements of the central points
of the rib elements

$\mathbf{H_u}$ n × m matrix of displacements of the central points
of the neck elements

$\mathbf{L_u}$ n × m × o matrix of displacements of the central points of
the enclosed air space elements elementmittelpunkte

$\mathbf{S_s}$ n × s matrix of string elements of a string s

$\mathbf{D_v}$ n × m matrix of displacements of the central points of the
top plate elements

$\mathbf{B_v}$ n × m matrix of velocities of the central
points of the back plate elements .

$\mathbf{S_v}$ n × s matrix of velocities of the string elements of a string s

$\mathbf{D_a}$ n × m matrix of accelerations of the central points of
the top plate elements

$\mathbf{B_a}$ n × m matrix of accelerations of the central points
of the back plate elements

$\mathbf{S_a}$ n × s matrix of accelerations of string elements of a string s

$\_\mathbf{D_a}$ intermediate memory of the spatial vectors of the central
points of the top plate elements

$\_\mathbf{B_a}$ intermediate memory of the spatial vectors of the central
points of the back plate elements

$\_\mathbf{S_a}$ intermediate memory of the spatial vectors of string
elements of a string s

$\mathbf{G_u}$ rib geometry vectors

Physical Parameter

$\mathbf{D_m}$ mass of each element (in this case in the top plate)
$\mathbf{D_\varrho}$ density of the wood (in this case in the top plate)
$\mathbf{D_{Ex}}$ Youngs modulus in the x direction (in this case in the top plate)
$\mathbf{D_{Ey}}$ Youngs modulus in the y direction (in this case in the top plate)
$\mathbf{D_h}$ height of each element (in this case in the top plate)

F	force
M	moment
$\mathbf{F}_{\mathbf{cross\ x\ y}}$	transverse force in the x and y directions
$\mathbf{D_v}$	velocity
$\mathbf{D_a}$	acceleration
v	velocity
a	acceleration
r	radius
δx	separation in the x direction between two neighboring elements
δy	separation in the y direction between two neighboring elements
δz	separation in the z direction between two neighboring elements

Guitar Parameter

ZH	rib height
LK	length of the body in the x direction
BK	width of the body in the y direction
MS	scale length
$SL_{\mathbf{u}}$	position of the sound hole
SL_r	radius of the sound hole
DD	average thickness of the top plate
DB	average thickness of the back plate
DZ	average thickness of the ribs
HB	width of the neck
HD	maximum thickness of the neck

Transformations

$\mathbf{e}_i$	base vectors of the guitar, so $\mathbf{u} = u_i \mathbf{e}_i$ ($\mathbf{u} = \sum_{i=1}^{3} u_i \mathbf{e}_i$)
$\mathbf{Z}^{axis}$	Base vectors of the rib elements
$\mathbf{T}^{i,j}$	Transformation matrix of the rib elements i, j

Time and Analysis

t	time
Δt	time interval
SR	sampling rate

f	frequency
ω	angular frequency $2\pi f$
$F()$	continuous amplitude spectrum
$F[]$	discrete amplitude spectrum
$Z()$	continuous time series
$Z[]$	discrete time series
z^x	Z transformation
CWT	continuous wavelet transformation
DWT	discrete wavelet transformation
FFT	Fast Fourier transformation

1

Introduction

1.1 General Remarks

This book describes a new paradigm in instrument acoustics – one based on time-dependent transient analysis and simulation of complete musical instruments. The prevailing view in instrument acoustics currently remains a static one. Instruments are described in terms of their static characteristic eigenfrequencies or their frequency-dependent properties concerning the dissipation of the sound. However, the music played on musical instruments is never static. On the one hand, amplitudes and/or pitches of the sounds radiating away from an instrument always change steadily, even during what is called the quasi steady-state. On the other hand, it must be borne in mind that musical instruments are systems of individual parts that are linked together, and so their properties are whole complexes that can only be understood through the interactive and coupling properties of the individual subsystems. This complexity of musical instruments cannot be regarded here purely as a combination of static properties of the subsystems; in reality the true vibrational behavior depends on the transient. Without this temporal dimension, investigation of how musical instruments produce their sounds would miss out important parts of the sound, and the instrument as a system cannot be understood in its entirety.

This transient synthesis is demonstrated here by computer-aided modeling using the guitar as an example. The modeling is carried out using the Finite-Difference-Method (FDM). The objective is to simulate the transient behavior of the instrument and to use the analysis to draw conclusions concerning basic properties of the guitar, and also concerning basic mechanical properties. The modeling system used proved to be appropriate for this context.

This investigation was also necessary for other reasons. In earlier works by the same author (Bader 2002a) the initial transients of guitars and other instruments were analyzed in detail using methods of signal processing, fractal dimension analysis and other related methods. However, using purely analytical methods to investigate sounds, it was not possible to clarify what the physical causes of these properties of the initial transients might be. To determine these causes, the concept of a complete simulation of a guitar on a computer appeared to be a suitable method for generating the sounds produced in the tones of guitars. The correctness of this simulation, which is monitored in its individual parameters and stages, allows the causes of the steady-state behavior to be determined with great accuracy. At this point it must be said that many of these causes would be difficult to recognize in an experimental situation, i.e. by manipulating a real instrument. An example of this would be the type of coupling existing between the top plate and the back plate of a guitar via the ribs. The bending wave in the top plate is transformed into a longitudinal wave in the ribs. When this longitudinal wave reaches the back plate, it becomes a transverse wave again. But waves within the guitar that do not have a transverse component cannot be measured directly just anywhere on the guitar body. And if, for example, we wanted to determine the corresponding longitudinal movement at one particular point in the top plate, we would have to saw it open. The chain of transfers outlined here, characterized by the coupling of transverse and longitudinal vibrations, is an important part of the sound of the guitar, as will be explained in more in detail in the chapter on the steady-state. It is this process that allows the ribs to dissipate most of the sound they receive from the top. The result is a type of initial transient with several stages, which initially begins with the attack in the top and which undergoes another change in its sound after about 5 ms. This process (amongst other things) gives the sound extra richness and makes the initial transient slower, and thus longer all of which are beneficial properties for a solo instrument played alone, as is often the case for the guitar. The ribs are a relevant aspect in both the construction and the sound, and go back to citoles in the Middle Ages, and perhaps even the chordophones of the Hethites, ca. 1000 BC, which are possibly the stringed musical instruments depicted on ancient pottery fragments. (For a detailed article on the development of the guitar see (Jahnel 1986)).

The interactions between the vibration in the top plate and that in the ribs brings about a heaviness of the sound that becomes a problem if the instrument plays a solo in a musical context. Two examples of how the rather "weighty" timbre of the concert guitar can be avoided are the flamenco guitar and the electric guitar. The electric guitars of the big-band era were acoustic or semi-acoustic instruments and as such still had ribs, and they sounded very full and muted owing to the relatively thick, polished strings prevailing at the time. However, the height of the ribs decreased progressively in the second half of the 20th century ("thinline" models) until they disappeared altogether from the electric guitar. This also meant that there was no air space any more, which also led to the disappearance of the additional "short, pumping sound" at the onset of the tone. Electric guitars are most commonly used in bands, where they are required to produce sounds within a clearly defined range of frequencies within the overall sound, namely the middle pitches. Any overlap with the frequency ranges of the other instruments, such as the bass or the percussion, often leads to an ill-defined, fuzzy sound. There are numerous musical tasks (principally as an accompaniment) where the electric guitar should provide "patches" or "dabs" of sound within an overall picture spanning a wide range of frequencies. Regarding aspects of sound composition, similar considerations apply to the flamenco guitar – virtually restricted to concerts – but during the period of its development in the nineteenth century it was one of the newest flamenco instruments (appearing later than castanets, for example) and could still be heard in peñ as and bars, where it is very loud. The flamenco guitar also has ribs, but its construction is much lighter overall. Although the effects that the construction and the materials used have on the sound of the flamenco guitar have not been examined in detail, the sound it produces is short and direct. This is probably also the reason for the frequent and rapid repetitions of notes in the rasguedo or tremolo techniques, which are intended to produce a sound which is as penetrating as possible without becoming "heavy".

A second motivation for this detailed investigation of the guitar has to do with questions and methods of observation in the scientific study of music. In musical acoustics, problems with instruments are often approached from a physical point of view whose relevance to actual music is not very obvious to musicians, instrument-makers or music scholars. Many studies of instrument acoustics still start from a physical standpoint, and often serve to confirm physical theories "using delightful music as an example". This state of affairs is characterized by the fact that such analysis of musical instruments hardly ever investigate real plucked, bowed or blown tones. It is much more often the case that the tone or sound which is analyzed in the laboratory is produced by machines playing the instruments (Hutchings 1997) Vol. I p. 499 (Galluzzo and Woodhouse 2003). This is certainly a welcome method, at least initially, for physical reasons. Standardized conditions have to be created in order to exclude interfering variables and to be sure which physical parameters lead to which alterations in the sound. However, after such investigations it is just as advisable to play real sounds and to examine whether the properties found during the physical studies are still relevant in a musical context. This may not be as interesting to physicists, because musical questions then come into play, which by contrast are central to the study of music. Conversely, scholars of music welcome the assistance of physics in clarifying musical, in particular acoustic, problems with sound. As an example in this book I will deal with the coupling between the strings and the top face of the guitar in detail. In the theoretical literature this is described as impedance problem with a first order differential equation. The resulting resonance spectrum, however, only correlated with a resonance spectrum created by artificial stimulation. Only this year has it been discovered that it does not have much in common with the resonance spectrum of a genuine tone played on an instrument at the fundamental pitch/fundamental frequency. The reason for this is explained in this book and a solution to the problem is offered.

We have now arrived at another problem that arises directly from the aspects mentioned above and that must be addressed by the study of music, namely time-dependent transient analysis. Music is always in flux. The spectral analyses of musical sounds show that they have quasi-steady-state, but not a true steady-state. And this is true even of a single played tone, even if it is generated mechanically by a machine and is therefore "artificial". It is easy to see how much more movement there is in the genuine musical playing of an instrument, considering temporal parameters relating to the phrasing and timbre. Transient analysis has been used on individual instruments (for a summery see (Reuter 1995)), but in such cases it has generally only been analytical determination of which special features occur in the instrument during the initial transient. There have not been any attempts at closed analytical or iterative solutions. And nearly all the Finite-Element-Analyses of instrument bodies were only carried out to determined their eigenvalues (see chapters on investigations of the guitar in this monograph). Transient analysis has two problems. To be able to provide a closed analytical solution at all, the bodies must be free from nonlinearities. In practice this is virtually never the case with real instruments. The closed iterative solution, however, requires a large amount of work, as can be seen later in this monograph. The whole geometry must be entered into the model and then every possible influence that any element of the body might have on any other must be investigated, in order to determine whether the calculation is plausible in these cases. But musical instruments are complicated constructions because this is the only way that they produce the richness of sound and the characteristics of a certain instrument – which we recognize. In the case of the guitar, for example, the sound typical of this family of instruments is usually easy for listeners to recognize, whatever the quality of the individual instrument or the strings used. Players and listeners, however, will be less interested in the "typical" sound of guitars in general than in the particular sound of an individual instrument, which can be described by its temporal and spectral characteristics. For every musician and instrument-maker, and for music studies, the audible and acoustic details are the interesting ones. This is what makes the music "musical", and this is also where personal expression and stylistic features come into play, this is where nuances differentiate the layman from the professional, and this is what we use in the final analysis to assess compositions and their performances. The principles of music reveal themselves in the details. Of course, we have to admit that these details have only become accessible to the extent they are researchable today, mostly owing to modern computers and software techniques of objective investigation which have only recently become available. Regarding the questions concerning instrument acoustics, it was also necessary to develop the theory of nonlinear dynamics and the coupling of complex systems (synergetics), to be able to carry out certain investigations.

Overall this work is an attempt to define more precisely the physical synthesis of sounds from instruments so that they can also be approached from a musicological point of view.

1.2 Brief Summary

First of all this monograph describes the current state of research into the guitar, and also concerning possible iterative procedures such as the Finite-Element-Method (FEM). To this end the special software packages MTMS, MSA and Wavelet were programmed and their mode of action is explained here, so that the reader will be familiar with them when they occur later in the text in the context of the results this software has produced. This is followed by a brief summary of the basic equations for the mechanics of vibrating bodies. I then deal with the difference between continual and discrete mechanics, elaborating on the fundamental differences not only of a technical nature, but also those concerned with views of space and time. There follows a description of the methods that have been developed especially for analyzing guitars. On the one hand these represent a further development of the Finite Difference Method (FDM), such as the coupling of longitudinal and transverse waves. Other innovations relate to the geometrical representation of possible curvatures created by a state of tension. These are initially shown as new differential equations, and then also made discrete. A new type of damping is presented, which does not alter the eigenvalues of the geometry and is therefore extremely useful for the case that interests us, namely the guitar. In addition, there is a theoretical explanation of the constant amplitude modulation, which accompanies almost all the overtones of nearly every instrument. The final part of my investigation describes three results from the calculation. First, the development of the initial transient is given an extremely detailed examination. The times taken for the initial transient to develop in the individual elements of the guitar are given. In addition, it is possible to allocate the sounds radiating from the various individual parts of the guitar to the overall sound. Secondly, the coupling between the strings and the top plate is investigated. The problem with the prevailing view of impedance of the guitar until now has been recognized by other researchers, namely a reduction of overtones; this problem is solved here. On contrast, an impedance theory of second order is introduced here as a solution to the problem, and on the other hand it is also required on the ground of physical considerations. This book concludes with a description of the finger noise made when the fingers slide over the strings before plucking. A theory of how this noise arises is put forward and it is shown that it can be used to obtain good results.

2

The Musical Transient-Modeling Software MTMS Developed for this Study

For this study a stand-alone software program was designed and written especially for the operating system WINDOWS. This program calculates transient behavior, that is the state of a system at successive points in time. This makes it Musical Transient-Modeling Software (MTMS). As a stand-alone application (SAA), on the one hand it allows detailed data input, realized using dialog boxes and check boxes. Data input is followed by calculation of the transients. These calculations are visualized as studies of the vibrations in slow-motion. This is carried out using various windows showing a visual 3-D representation of the individual parts of the guitar, such as the plates or other components. Finally, after the time series are calculated they are exported for further processing by time series analysis programs.

Several independent WINDOWS programs were also developed for analyzing these time series, and they are described below. A commercial version of the MTMS program is nearing completion. This will provide instrument-makers, instrumentalists and teachers who are not experienced programmers with a simple and flexible means of trying out various settings of the guitar parameters and hearing the calculated resulting sounds. As this software is used in the following chapters of this book, a brief description will be given here. Later I will be referring to the relevant parts of the program used for calculating the results and for the visualizations.

The programming languages used were FORTRAN, C++, C#, PASCAL and native machine code (assembler). In addition, .dll files (drivers) were programmed, which were flexibly loaded and controlled by the programs. The visualization was done using either WINDOWS CLR or a DIRECT-X representation. Object-oriented programming (OOP) allows any extensions of the code, to include other musical instruments and also other geometries where calculation of the transient sounds is desired. All the programming and associated work was carried out by the author of this book independently and alone. No other external programs were used in addition, and no work or suggestions from other researchers or programmers was used in the realization.

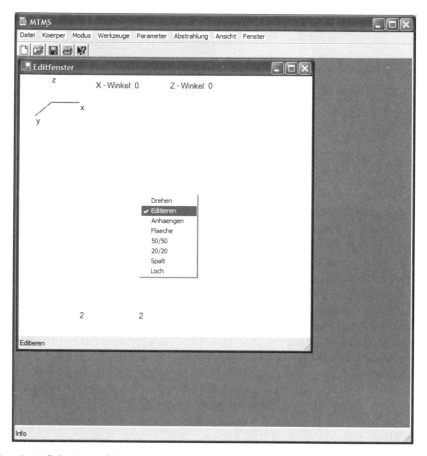

Fig. 2.1. Selection of the plate to be calculated, in edit mode for the plate in the MTMS program

This was necessary for reasons of academic propriety, and also to avoid being dependent on third parties where commercial use is concerned.

MTMS currently consists of two major parts. The first part can calculate a plate with any geometry (see Fig. 2.1 and Fig 2.2 for the material dialog box for plates). The other implements a complete guitar. Work is also continuing at present on a snare-drum. Furthermore, a method is already being tested which should make it easy to use FDM directly on any geometry, by means of a meshing process, in a similar fashion to its use in FEM. Here it is possible to have any number of grid elements along with an adaptive grid generation, which allows increases in the number of grid elements used in regions of severe load or deformation, similar to FEM. This means that an additional module will be created, allowing the users to enter any geometries – and thus their own instruments – and to calculate transients.

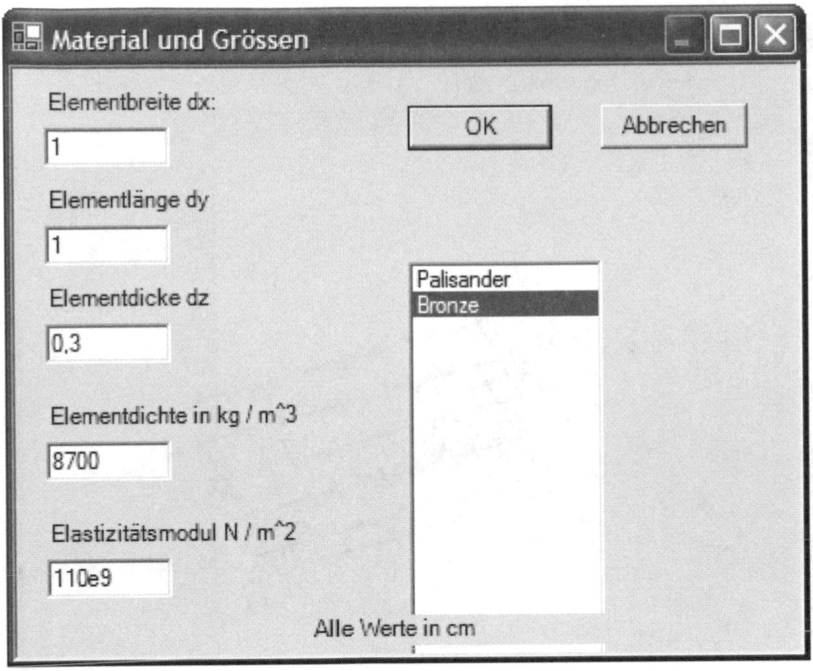

Fig. 2.2. Material box for the plate in the MTMS program

2.1 Part of the MTMS Program for Plates

The construction of a plate that can be calculated using MTMS is carried out by means of an editor; a keyboard is used to add the individual elements to the plate. This can be done by selecting appropriate parameters in the dialog box appearing upon a right mouse click for individual elements and also for series of elements, as required. Moreover, it is possible to select plates that have already been prepared. These could be plates of various different sizes, or a plate with a hole in the middle (for representing the calculation of waves in the course of the transient), or plates with a slit or a double slit (see Fig. 2.4).

Then any plate can be constructed. The example in Fig. 2.3 shows a plate as a grid, very simply constructed in individual element mode. However, here too, it is possible to use existing structures as presets, as described above.

Finally, the plate can be calculated. There are several ways of doing this. First, the plate can be stimulated with a certain sine wave. The plate vibration is then shown in slow motion. At this stage three different types of visualization are possible. The type to be selected depends on the nature of the problem. Interpolated plane representations demonstrate the course of a wave more precisely, while a grid-type representation results in a more detailed description of the plate motion. Of course, the output also contains the exact values, including data in the form of ASCII files or binary files, and these

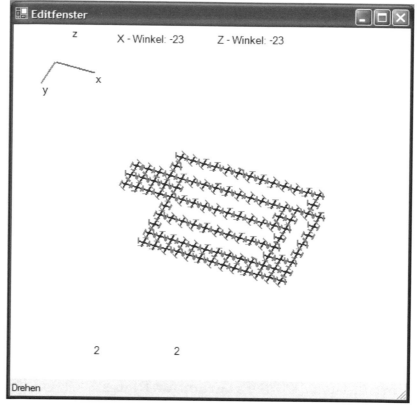

Fig. 2.3. Construction of a grid plate in edit mode in the plate part of the MTMS program

values can then be further processed by other programs. Work is continuing on the integration of a separate signal-processing part to the program.

2.2 MTMS Software for Guitar Parts

The second geometry implemented in the MTMS software is that of the guitar. In this part all the default values are set at the beginning in such a way that calling up the guitar calculation menu entry is sufficient to set all the parameters to values typical of guitars right at the start. Of course these values can be changed. For each part of the guitar there is a dialog box for altering the material values, damping values and other values. The menus in the top panel of the material box of Fig. 2.5 refer both to the so-called influence dialog box (see below) and to a dialog box that wholly or partly resets the displacements and/or velocities of the geometry. For example, when the guitar is vibrating, damping of one part of the instrument can be simulated, e.g. the

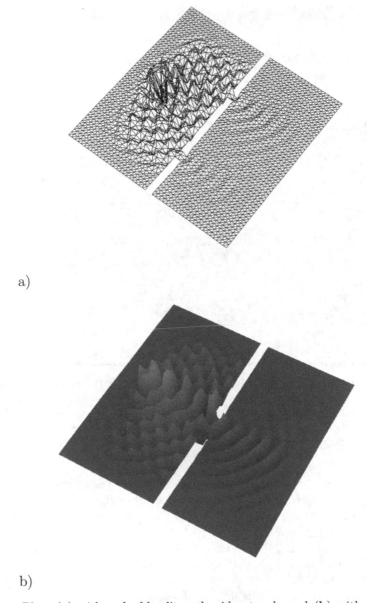

a)

b)

Fig. 2.4. Plate (**a**) with a double slit and grid network, and (**b**) with a simple slit and interpolated plane representation in the MTMS program (snapshot). The frequency and amplitude of actuation were chosen to suit the geometries. In this case the boundaries of the plate are fully damped to prevent reflection. This is necessary for the double-slit experiment, to exclude reflections which could cause interference. The changes to these boundary conditions were made in the program code itself; the automatic integration of the revised program code is a feature that will be added at a later date

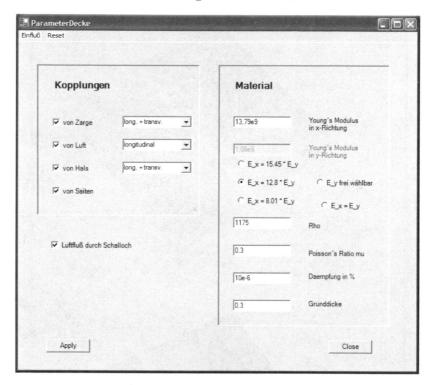

Fig. 2.5. Dialog box in the guitar part of the MTMS software for determining the material values and the damping, the coupling properties and other similar values. Here the dialog for the top plate is shown as an example. There are similar boxes for all the other parts of the guitar, such as back plate, ribs, neck, enclosed air space, strings and global values

damping caused by laying a hand on the body of the instrument – or more generally on the string – in order to stop it vibrating. All of these measures can be programmed, and applied in a flexible manner at any time while the guitar is vibrating (see Fig. 2.7 for global and coupling parameters).

Furthermore, Fig. 2.6 shows the dialog box for possible influences on the guitar. Here Dirac impulses at a point t_0 in time can be added at a grid point P_0 of the structure in its z-direction Z_{P_0} by adding to the existing z-displacement z_{P_0} in the calculation of the guitar structure (R) an impulse with an amplitude α like

$$z_{P_0}(t - t_0) = z_{P_0}^R(t - t_0) + \alpha \, \delta(t - t_0) \tag{2.1}$$

with the delta-function

$$\delta(t - t_0) \begin{cases} 1 & \text{for } t = t_0 \, ; \\ 0 & \text{for } t \neq t_0 \, . \end{cases} \tag{2.2}$$

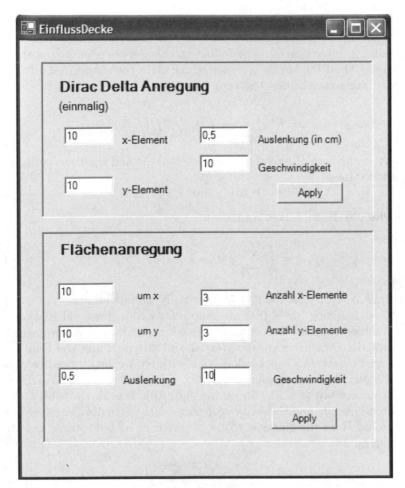

Fig. 2.6. Dialog box in the guitar part of the MTMS program for determining the influence to be exerted on each part of the guitar – here the top plate. Dirac impulses and area impulses can be programmed at this point. The Dirac impulse relates to an element and is determined in relation to the displacement and the initial velocity of the element selected. The area impulse can be regarded as a type of golpe, i.e. a knock on the guitar's top plate. Here the central element and the number of elements surrounding this central element that the impulse should trigger can be determined. The influence can also be implemented during the ongoing calculation. Once the parameter has been set using the Apply button, the impulse is applied to the guitar, which is already in motion

Furthermore, area impulses can be added to the different guitar parts,

$$z_{P_{x,y}}(t - t_0) = z_{x,y}^R(t - t_0) + \alpha \, \delta(t - t_0) \tag{2.3}$$

with $P_{x,y}$ being the used grid points, where α can also be a function, i.e. a three-dimensional gauss-function. This method is implemented to calculate eigenvalues. For this, in the material-dialogbox of the guitar parts, i.e. the top plate is decoupled from the other parts of the guitar geometry. On this part an impulse is programmed. The following calculation results in a transient time series. The FFT of this time series calculate the eigenfrequencies of the geometry using the Fourier-Theorem

$$F(\omega) = \frac{1}{2\pi} \int_{t_0}^{t} e^{-i\ \omega t} f(t)\ dt \qquad (2.4)$$

and here in the case of a discrete time interval Δt and the time vector of the amplitude values

$$f[t_0 + k\ \Delta t] \quad \text{with } k = 1, 2, 3 \ldots \qquad (2.5)$$

by the discrete Fourier-Transformation

$$F[\omega] = \frac{1}{N} \sum_{k=1}^{k=K} f[t_0 + k\ \Delta t]\ e^{-i\ \omega(t_0 + k \Delta t)} \qquad (2.6)$$

in the time interval $t_0 < t < t_0 + k\ \Delta t$ with N sample points.

Two more dialog boxes that are required for the guitar calculation, refer to the representational and working behavior and to global parameters. At this point the simulation can be started and stopped, and the Play button can be used to obtain a quick calculation without a visualization, calculating a certain period of time (again controlled via a dialog box) and writing the sounds thus created to a .wav file on the hard disk. It is also possible to set the sampling frequency and the zoom properties of the visualizations (see below). If one point P_0 of the guitar is to be actuated in its z-direction z_{P_0} using a sine wave f_0, as

$$z_{P_0}(t + t_0) = z_{P_0}^{R}(t + t_0) + \alpha\ \sin(2\pi f_0 t + \Phi)\ , \qquad (2.7)$$

the frequency f_0 and amplitude α can be set here. In addition, longitudinal and transverse vibrations can be switched on and off globally.

The string dialog box governs the displacements of the strings (see Fig. 2.8). At the beginning of the calculation it is determined here which strings are displaced, at which positions, and in which manner. This can be programmed either as a trapezoidal function of the initial z-displacement $z_S(n)$ of the n-th string grid point S with length l, a maximum amplitude α and a plucking point x as

$$z_S(n) = \begin{cases} \alpha\ \frac{n}{x} & \text{for } 0 < n < x\ ; \\ \alpha\ \frac{l-n}{l-x} & \text{for } x < n < l \end{cases} \qquad (2.8)$$

On the other side, this can also be programmed as an impulse in a way described above with the guitar geometry. Naturally, any additional displacement can be added here at any time. Following this method, whole guitar pieces can be programmed. Furthermore, the individual tones or chords can

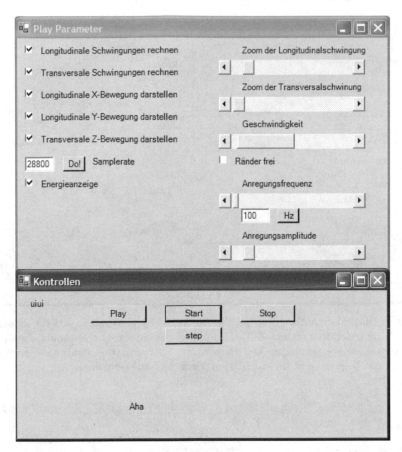

Fig. 2.7. Dialog boxes in the guitar part of the MTMS program, which set the control of the development of the sound and global parameters for the calculation

be programmed as displacements of the strings as a function of time. It is also possible to damp the strings (see above). Both the position at which the string is plucked and the volume of the sound can be adjusted. The exact method of plucking, which has to be programmed to get most realistic sounds is also influenced by the naturally appearing finger pre-scratch noise of the real sound. Its exact composition and its synthesis are discussed in the chapter on the results from the guitar calculation at the end of this monograph.

The radiation from individual elements within the parts of the guitar can be illustrated "live", that is in real time, during the calculation. Figure 2.9 shows the element (5,5) in the top, for example. At present work is progressing on a Fast-Fourier Transformation (FFT) in real time. This enables parts of the overall sound contributed by the individual parts of the guitar to be followed "live".

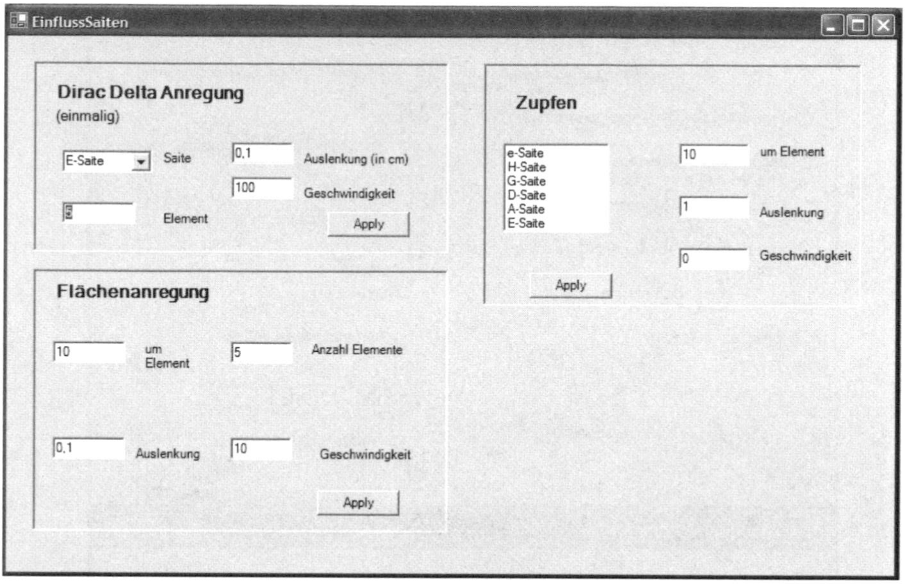

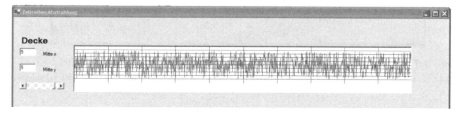

Fig. 2.8. Dialog box in the guitar part of the MTMS program that sets the displacement of the individual strings. Either individual impulses or trapezoidal functions can be used for the displacements. Here again, this actuation can be set to occur either at the beginning of the calculation, or at any time during it

Fig. 2.9. Time series of the radiation from one element of the guitar, here one of the elements in the top plate. This analysis box shows the radiated time series of only one element, in real time

In the dialog box for the matrices the values of the individual guitar elements can be shown as ASCII codes either in an independent WINDOW or written to files on the hard disk.

Finally, I wish to present an example of a visualization. Since all parts of the guitar are shown in separate windows, and can be rotated or manipulated using various displacement zooms, here I shall only provide as an illustration an example of the representation of the top, ribs, strings and a complete view. These representations can be enlarged or reduced as desired. In addition, it is possible to visualize the internal air space and the neck. In the simulation you

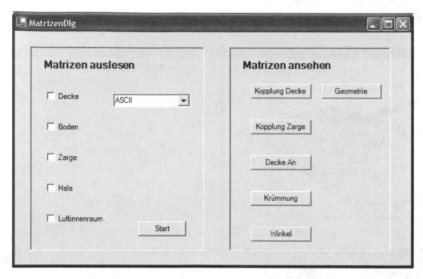

Fig. 2.10. This matrix dialog is used for the entry or output of matrix values

can see the various parts vibrating in slow motion, and so can also observe the influence of each part upon neighboring parts. The types of representations can be varied a great deal. For example, it is possible to have a display of only the longitudinal movements or only the transverse movements, or any combination of these two types of vibrations. Alternatively it is possible, for example, to zoom in on a particular place in the enclosed air space of the guitar in the complete view, that is, to "look inside" the guitar. In doing so, the top and back plates and the ribs can be observed in motion at the same time, which is necessary for understanding the spread of wave impulses through the body of the guitar, for example.

2.3 WAVELET Program for Calculating a Wavelet Transformation, the Spectral Centroid and the Spectral Density

The WAVELET software for calculating a wavelet transformation from a .wav file was also realized for this book as a stand-alone application (SAA) for WINDOWS. The temporal development of the spectral centroid of the sound and the temporal development of the spectral density are calculated from the values of the wavelet transformation. These two indicators are very useful for estimating the psycho-acoustic perception of the brightness and density of the sound. In this monograph the calculation is carried out for various sounds, principally those of the individual parts of the guitar. The parameters used are given in the relevant chapter.

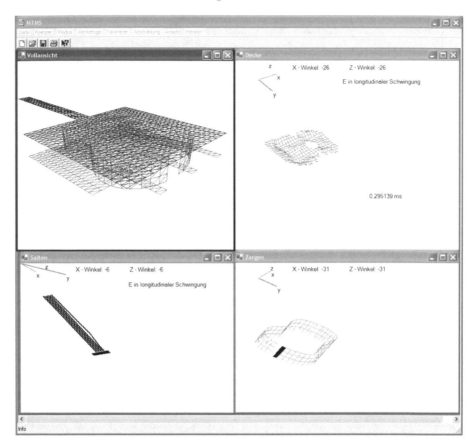

Fig. 2.11. Example of a visualization of top plate, ribs, strings and complete view of the guitar, using the MTMS program

2.4 The MAS Program as a Signal-Processing Tool for Music

In addition to the above, the author developed the musical analysis software (MAS) that was used for analyzing the calculated sounds in this book. This program unites various signal-processing analysis tools that have proved to be useful in relation to musical analysis, and especially of transient behavior. Figure 2.12 shows a time series (bottom) and its wavelet transformation (top). In its discrete form this is (Haase et al. 2002)

$$F[\omega, t] = \frac{1}{N} \sum_{r_{\min}=t-\pi w/\omega}^{r_{\max}=t+\pi w/\omega} f[(r_{\min} + k\Delta t)s] \ e^{i\omega t} \ \beta(r_{\min} + k\Delta t)$$

$$\text{for } k = 1, 2, 3 \ldots 2\pi s w/\omega. \tag{2.9}$$

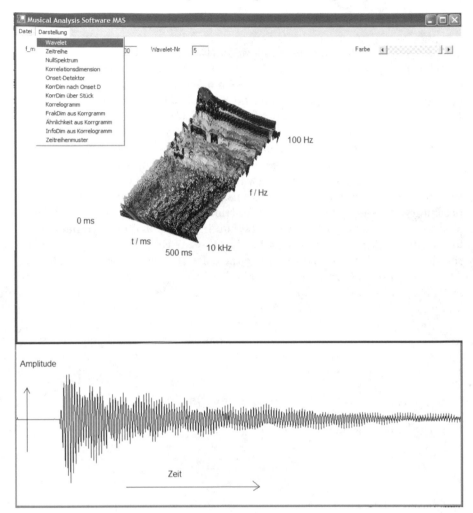

Fig. 2.12. Examples of the calculation possibilities in the musical analysis program MAS. For more possibilities, see text

Here $F[\omega, t]$ is the Wavelet-transformed function of the frequencies $\omega = \{\omega_1, \omega_2, \omega_3 \ldots\}$, and the time points $t = \{t_0, t_1, t_2, \ldots\}$, where both can be chosen arbitrarily. $F[\omega, t]$ is calculated out of the discrete time series $f[l]$ with $l = 1, 2, 3 \ldots s \, t_{\max}$ with s being the samplerate and $t_{\max}$ being the length of the soundfile in seconds divided by the number of used sample points N. Each frequency ω has the boundaries $t \pm \pi w/\omega$ with the Wavelet number w, so in the convolution with the time-amplitude series it has its own window. Larger Wavelet numbers improve the frequency resolution, smaller values improve temporal resolution of the transform. This function is additionally windowed like with a normal Fourier-Transformation through a gauss-function

$$\beta(r_{\min} + k\Delta t) = e^{-(r_{\min} + k\Delta t)^2/c} , \qquad (2.10)$$

with reasonable values for the window width c.

The term $f[(r_{\min} + k\Delta t)s]$ holds for the k-th sample point with the frequency $\Delta t = 1/s$; the term, beginning from the point $r_{\min}$ in time, depends on the calculation time point t and on the calculation frequency ω. We see with the multiplication of the sample points with the expression $e^{i\omega t}$ that the Wavelet transform used here is a discrete form of a complex Morlet-Wavelet.

This transformation can be rotated in 3-D as desired in real time. In many cases this is very helpful. For example, it is often difficult to clarify in a static view of the overtone development, whether the fluctuations seen relate only to frequency or to amplitude fluctuations, or to a combination of the two. The possibility of zooming in on the wavelet transformation in real time in order to observe the minutest parts of the spectrum in detail, is often extremely useful for clarifying such questions. In addition, the coloring can be adjusted using the slider control (top right) to change the color thresholds for the amplitudes. This enables regions with the same amplitude to be determined very exactly and easily. It is also a simple matter to change the frequency band of the analysis and of the wavelet number.

The MAS program has a range of further analysis tools, such as calculating a "zero spectrum", that is, a zero crossing detector, which calculates the frequencies of sine waves as well as of vibrations with an enriched overtone structure and their time series like impulse or saw tooth series with

$$f = 2/(n\Delta t \Theta(g_1)\Theta(g_2)) \quad \text{for } \Theta(g_1) = \Theta(g_2) = 1 \qquad (2.11)$$

with the heavyside-function for which holds

$$\Theta(\text{true}) = 1 \qquad (2.12)$$
$$\Theta(\text{false}) = 0 \qquad (2.13)$$

and the inequalities

$$g_1 = \text{Sign}(f[t]) \neq \text{Sign}(f[t + \Delta t]) \qquad (2.14)$$

and

$$g_2 = \text{Sign}(f[t + n\Delta t]) \neq \text{Sign}(f[t + (n+1)\Delta t]) \qquad (2.15)$$

with

$$n = 1, 2, 3 \ldots \quad \text{then also } n > 0 \qquad (2.16)$$

and the requirement

$$n \text{ be minimal} . \qquad (2.17)$$

Here we assume the time series having positive and negative amplitude values and being steady. I.e. a sine tone crosses the zero aches twice each period. Is the zero crossing detected happening between two successive time points t and $t + n\Delta t$ with the sample frequency Δt as the smallest time interval, then it holds $f = 2/(n\Delta t)$, because of $f = 1/T$ as described in (2.11). This equation also includes two condition functions, described here as Theta-functions. $\Theta = 1$ holds, if the functional () of Θ is true, $\Theta = 0$ holds, if the functional of wrong. So here as functionals the conditions for the time points of the zero crossings are used. Here it holds, that the function f of the discrete time-amplitude series experiences a change of sign, also g_1 and g_2 respectively hold. In g_2 the distance $n\Delta t$ is already included. With the requirements (2.14) and (2.15) the minimal temporal distance of two zero crossings is detected, if condition (2.17) is considered. As this time interval is half a period, as mentioned above, the frequency is calculated through this method. The zero crossing detector can play a role in analysis of the vibration of guitar strings, because it finds out the fundamental frequency without costly use of Fourier- or Wavelet-transformations by extracting the fundamental without calculation of the spectrum directly out of the time signal. Of course, this only works, if the basic pattern of vibration if known, like is the case with the trapezoidal displaced string after plucking.

Additionally, in Wavelet-Software an automatic calculation of a fractal correlation dimension is added. We do not use it in this work, so it is not described in detail here.

Another useful tool is the onset-detector for music pieces, which calculates the times of the onsets of the guitar tones and so provides important information for the analysis of crucial parts of the time series. The data of the onset time points can be used in the Software i.e. to calculate at precisely this points of interest in terms of increased complexity of the musical piece automatically a Wavelet-Transform of a fractal correlation dimension.

Also useful for the analysis of very complex sounds with several separate overtone structures is the correlogram. With this method, the correlogram data can be used to perform an similarity calculation to separate different parts of musical pieces, like verse of refrain and to calculate the similarities between them. Also the calculation of a fractal information dimension out of the correlogram is possible, where additional information about the complexity of musical pieces or single sounds can be found. The time series pattern is a related method, which detects time series periodicities out of the analysis of musical pieces. As the use of these analyzing tools is beyond the scope of this book, here we point to the literature (Bader 2002a).

Finally we use the spectral centroid in the further discussion as well as the spectral density. The spectral centroid Z is a psychoacoustic measure of the brightness of sounds, which comes out of the spectrum of the several partials and their amplitudes as

$$Z = \frac{\sum_f f\, A(f)}{\sum_f A(f)} \tag{2.18}$$

with $A(f)$ being the amplitudes of the frequencies f. The spectral centroid is useful in the analysis of the guitar back plate and the ribs to find out, if the curvature of these parts increase the brightness of the spectrum, as claimed by guitar builders.

The spectral density D is calculated similar to the Kolmogorov-Sinai entropy (Argyris 1995) as

$$D = \frac{1}{N_f} \sum_{N_f} A^{\text{normal}}(f) \, \ln(A^{\text{normal}}(f)) \,, \tag{2.19}$$

with N_f being the amount of spectral components. Here, the normalized amplitudes $A^{\text{normal}}(f)$ are used with

$$A^{\text{normal}}(f) = \frac{A(f)}{\sum_f A(f)} \,. \tag{2.20}$$

So it is

$$\sum_f A^{\text{normal}}(f) = 1 \,. \tag{2.21}$$

The spectral density can then have the values

$$0 \leq D \geq 1 \,, \tag{2.22}$$

where $D = 1$ only is reached, if the amplitudes of the components in the used frequency band are the same, which is just the case in white noise. $D = 0$ is the case with a spectrum of just one frequency component. So the measure of the spectral density is suited to represent the density of a sound, where high values are associated with a large amount of components and so as dense, and low values are associated with a "thin" sound. The spectral density is used in this work to describe the amount of noise in the different guitar parts, as high values are mostly caused by additional frequencies next to the harmonic spectrum.

3

Continuous and Discrete Mechanics

3.1 Overview

The guitar calculated here is a mechanical system with a complex geometry. The calculation of its vibrational behavior is therefore so complicated that an analytical solution is no longer possible in the sense of a continuous mechanical system. Therefore a discrete approach was chosen: the Finite-Difference Method (FDM) (Haile 1997; Tolonen 1998). This approach is applied in a series of scientific areas such as nonlinear mechanics (Zimmermann 1985) and geology (Bohlen 1998).

The FEM (Finite-Element Method), also used in this book, works with global functions (for standard texts to FEM see (Knothe and Wessels 1999; Hughes 1987; Bathe 2002). The global equations of virtual work is wanted here (see further discussion). As these global approach determines the global behaviour of the system, some overall laws are assumed here. This is not the case with the Finite-Difference Method, as here just basic mechanical laws are applied to a system locally, so the overall behaviour is a cause of these local descriptions. In time-stepping formulations, FEM and FDM may be nearly the same depending on the algorithm used.

3.1.1 Strong Formulation

A fundamental distinction can be made between a strong formulation and a weak formulation of mechanical problems. A strong formulation is the one used for closed analytical solutions. It does not contain any additional terms that vary with time, or other similar items. Of course it can also be solved discretely. The weak formulation, with any number of functions added to the equation, can only reach a closed solution in exceptional cases.

The differential equation of a vibrating string can be regarded as a simple example of the strong formulation. If

$$w_{,,x} = \frac{T}{\mu} \, w_{,,t} \tag{3.1}$$

with

$$w(0) = 0 \quad \text{and} \quad w(l) = 0$$

is the differential equation of the string with transverse displacement w, and its derivation with respect to space x and time t, the length l, the tension T, the linear density μ with the boundary conditions, of the strings being fixed at both ends $w = 0$ and $w = l$, then the solution

$$w(x,t) = \sum_{k=1}^{\infty} A \sin\left(k\frac{\omega}{c}(ct - x)\right) + B \cos\left(k\frac{\omega}{c}(ct - x)\right)$$

$$+ C \sin\left(k\frac{\omega}{c}(ct + x)\right) + D \cos\left(k\frac{\omega}{c}(ct + x)\right) \tag{3.2}$$

is analytically possible. The equation for the solution was guessed and found to be the solution of the differential equation. One can also say that the strong formulation is an expression using a coefficient. In the differential equation of the string, the coefficients can be altered as desired, but the basic model of the string remains unchanged.

3.1.2 Weak Formulation

The weak formulation always occurs when the system of equations is so complicated that either an equation for a solution can only be guessed with difficulty (i.e. it would be so complicated that there would be no use in having it), or is impossible. The latter case may occur with self-referential iteration systems, for example. If every successive solution requires and builds upon the previous one, and the system possesses nonlinear relationships, it is often the case that the results can no longer be predicted analytically. Our guitar can only be analyzed using the weak formulation.

The possible solutions to such problems will be discussed below. Although the calculation for the guitar presented here is dependent on time, and is therefore an iterative calculation, the discussion will first turn to the action principle as the basis for the following calculation. It will be shown how the equation of motion is derived from a simple mechanical view of energy. Conversely, it can be concluded from this that with temporal and spatial iterations of simple mechanical formulas, the total energy of the system must remain constant.

3.1.3 Action Principle

In this section the simple mechanical law $F = m/a$ will be derived from the Lagrange function of kinetic and potential energy. This is called the variation or action principle, and is also found in static problems using the Finite-Element Method (FEM) (Knothe and Wessels 1999; Hughes 1987). There the system of equations is determined from the variation of the potential energy of the system, and the deformation of the body is the solution to this system of equations. However in general, an iterative procedure that uses the Newton method for error minimization leads more quickly to the goal than does the solution of the system of equations for the body in a discrete formulation. However, in an FEM calculation, problems that are time-dependent are no longer solved using variation, but instead are mostly solved by applying the Newmark procedure (Hughes 1987; Wriggers 2001). A comprehensive description of this technique is given later in this monograph, but first the procedure of the action principle will be presented. One should note that each problem solved using the strong formulation can also be tackled by calculating the variation. These two procedures are equivalent.

The calculation of variation is one of the most feted proofs in physics. Feynmans path integral in quantum mechanics is one variation, which describes the movement of quanta. In general the procedure is applied when an equation for the movement of the system is to be derived from a view of energy. This is of interest in cosmology. Since in this field, the Lagrange function is the only way, it is very easy to determine the movement of particles along the world lines from it. But the calculation of variation is also used in string theory to provide the equations describing movement in complicated particle theories.

Here I shall demonstrate the basic principle, i.e. that every movement follows the principle of least action. Newtons law of motion will be proved. In addition, the action S is introduced, which is the temporal integral over the Lagrange function $L = T - K$, where T represents the kinetic energy and K the potential energy.

$$S = \int_{t_1}^{t_2} T - K \; dt \qquad (3.3)$$

If we now replace T and K with the expressions of energy, we obtain

$$S = \int_{t_1}^{t_2} \left(\frac{1}{2} m \left(\frac{dx}{dt} \right)^2 - mgx \right) dt \qquad (3.4)$$

We can imagine a ball, which has a certain kinetic and potential energy at each point along its trajectory. At each point in time, the potential energy is subtracted from the kinetic energy, and the result is integrated over the time taken. This integration is the resulting action S.

We now want to vary the expression. To do this, we will imagine a slight change to the perfect trajectory. This change has to be added at every point in time. We now obtain

$$S = \int (T(x + \delta) - K(x + \delta)) \; dt \qquad (3.5)$$

We think that out of all possible paths that the ball can follow between t_1 and t_2, the correct one is the one that with the minimum S, i.e. when the action is at its smallest.

If we did not integrate over time, the solution would simply be the disappearance of the functions derivative. But we must first integrate over a possible path and compare its effect with all other integrals. It is very helpful for us that when we are close to the right path, any small deviation from this path can only be a change of the first order at the most, which is true of every extreme point (and, strictly speaking, for a maximum value, too). We shall write out $T(x + \delta)$

$$\left(\frac{dx}{dt}\right)^2 + 2\frac{dx}{dt}\frac{d\delta}{dt} + \left(\frac{d\delta}{dt}\right)^2 , \tag{3.6}$$

and $K(x + \delta)$ as a Taylor series, so $K(x) + \delta K'(x) + terms\ of\ higher\ order$, and the only remaining term is dS, the variation in action:

$$dS = \int \left(m\frac{dx}{dt}\frac{d\delta}{dt} - \delta K'(x)\right) dt \tag{3.7}$$

dS should be as small a possible, i.e. zero. The problem is that δ appears both in its basic form and as a temporal derivative. Since these terms are dependent on each other, we cannot simply set δ to zero. We have to create an expression that is $\delta\ times\ (\ldots)$, so that dS can always be zero and δ can have any value, if we set the bracketed term "$(\ldots)$" to zero. The principle of calculating variation involves achieving an appropriate expression $\delta\ times\ (\ldots)$, and then setting $(\ldots) = 0$. The principle of the least potential action, which occurs in the Finite-Element Method, functions analogously. In a static analysis there are no temporal derivatives, i.e. there is no kinetic energy, but as we will see, the state of displacement is correct where the potential energy is at a minimum. And there, too, the energy equation must be reformulated so that the expression $\delta\ times\ (\ldots)$ is left over. The resulting equation $(\ldots) = 0$ can then be solved either by a Gaussian elimination procedure, or iteratively by error minimization (usually using the Newton method).

To achieve the expression $\delta\ times\ (\ldots)$ with our action dS, we have to do a partial integration. Since in general

$$\frac{d\delta}{dt}g = \delta\frac{dg}{dt} + g\frac{d\delta}{dt} \tag{3.8}$$

applies, then in our case for $g = m\frac{dx}{dt}$ we have either

$$dS = m\frac{dx}{dt}\delta\Big|_{t1}^{t2} - \int_{t1}^{t2} m\frac{d}{dt}\left(\frac{dx}{dt}\right)\delta\ dt - \int_{t1}^{t2} K'(x)\ \delta\ dt \tag{3.9}$$

or

$$dS = \int_{t1}^{t2} \left(-m\frac{d^2x}{dt^2} - K'(x)\right)\delta(t)\ dt . \tag{3.10}$$

Now we have arrived at the point we were trying to reach. We have an expression, which when multiplied with the variation δ, produces the change in action. But we want this to be as small as possible, i.e. zero. In order to have the smallest action for all values of δ, the expression in front of it must be zero, so:

$$-m\frac{d^2x}{dt^2} - K'(x) = 0 \ . \tag{3.11}$$

But this is nothing other than Newtons law of motion. So we have proved that the motion of a body corresponds to the principle of least action. This is the underlying principle behind the Finite-Element Method, but not the one behind the Finite-Difference Method. The latter is a method describing transient behavior, i.e. a development over time. So here of course a principle of action exists, but is no longer practicable. The variation just demonstrated above does apply over time, but only for one particle. The variation calculation in the FEM with potential energy applies to the whole body, but only at one point in time. The temporal motion of a body existing in space must also be subject to the principle of least action. If this motion can be described with simple differential equations, then the law of motion can also be derived from the effect. However, this is not practicable from one time interval to the next. The variation would have to be carried out across time *and* space. So we would obtain a system of equations that included all elements in space at all points in time, and would have to be solved for all of these. But as we have already seen, even with the static FEM a great amount of work is involved in solving a system of equations even only for one spatial deformation. So FEM analysis describing transient behavior are no longer calculated according to the principle of least action, but instead according to the Newmark procedure, whose special case of the central differences corresponds more or less to the principle of finite differences. The advantage of the FEM is that the energy conservation law automatically applies, since we are assuming an action equation that is an energy equation. With the FDM, and with the transient FEM, this conservation of energy is no longer taken into account in the extreme equation conditions, but is only present implicitly. The stability analysis for the FDM is, however, a prerequisite for the procedure to function at all. The view of energy that we take below is of interest when using the FDM, because a "thermal fluctuation" is found in experiments to belong to the system, which is also automatically contained in the FDM.

Now that the calculation of variation has been presented, we want to turn our attention to the basic principles of continuum mechanics, so that we can discretize them in the course of the discussion and to make them independent of time.

3.2 Overview of Continuum Mechanics

In this chapter we want to look at the basic principles of continuum mechanics, as they are important for our purposes. Discussions of thermal aspects will not be included, since they only play a subordinate role in the modeling of guitars. In addition, the terminology introduced in this monograph will be presented here, along with some basic definitions.

3.2.1 Definition of the Variables

The guitar is a mechanical system. We therefore speak of plates: the guitars top plate and back plate, ribs, etc. These plates are continua, occupying a region within space, and at every location they have a certain displacement, velocity, etc., and there are forces, moments and similar items, that affect them.

The plates are generally given the symbol Ω, and their boundaries are given the symbol Γ. Boundary conditions such as whether the plates are fixed or can vibrate freely are given the symbol R_Γ. So for the top plate, for example, we get:

$$\Omega \in \Re^3 \quad \text{where for } R_\Gamma \quad \text{it holds :} \quad \mathbf{u} = 0 \tag{3.12}$$

So the plate is a three-dimensional continuum in space, and anywhere on its boundary the displacement $\mathbf{u}$ is zero.

$\mathbf{u}$ is a vector, just as all letters in bold type denote vectors, matrices or tensors. So $\mathbf{u}$ encompasses all the three dimensions in space: the x, y and z dimensions. The individual displacements in these dimensions are denoted by r, s and t, or in the component notation by u_x, u_y and u_z.

Derivatives are written with a comma. For example:

$$\frac{d\mathbf{u}}{dt} = \mathbf{u}_{,t} \tag{3.13}$$

or

$$\frac{d^2\mathbf{u}}{dt^2} = \mathbf{u}_{,,t} \tag{3.14}$$

Velocities are given the letter $\mathbf{v}$ and accelerations the letter $\mathbf{va}$. The individual plates will be introduced later in discrete cases as matrices. They are then called $\mathbf{D}$ for the top plate and $\mathbf{B}$ for the back plate, for example. Since the software for the guitar, and thus all the important results, are discrete, these variables are not used in the continuous case. However, later we have $\mathbf{D_v}$ for the velocity of the top plate elements, and $\mathbf{D_a}$ for their acceleration, and correspondingly $\mathbf{D_u}$ for their displacement. In general we speak of $\Omega_{\mathbf{u}}$, for example, for the displacement.

Displacements lead to strains. These strains are denoted by ϵ and are also given as relative change in length, $\frac{\triangle\mathbf{u}}{\mathbf{u}}$. The resulting cross-sectional forces (stress) σ are F/A, i.e. force divided by area for the plate, and F/V, i.e. force divided by volume, for volumes. The external pressure is given the symbol p and has the same units as the cross-sectional forces of the plate.

The physical parameters are the elastic modulus E, which has two different values for wood, one along the grain and one across the grain, which are here always given as E_x for the modulus along the grain and E_y for the modulus across the grain, and together as the vector $\mathbf{E} = \{E_x, E_y\}$. The only three-dimensional case in the guitar is the air space. Since air is homogeneous, we only need one E modulus.

The mass is m, the density ϱ, and the height of the plates (i.e. their thickness) is h. The latter can in fact vary, but it can be calculated for any variations in height in the discrete case.

3.2.2 Mechanical Laws

In this monograph all parts of the guitar are represented as plates. The only exception is the internal air space, which is three-dimensional and does not exhibit any bending motion. The reason for this is that air cannot have any moments of twisting, since air molecules move past one another freely. But of course the air can be compressed in the transverse direction. So, since the air is a simplified form of the plate properties, the only equations used here are those for the plate and not those for a three-dimensional continuum. The air space is dealt with in the discrete case, as it only has differential equations of the second order, and can therefore be derived as an extension of the longitudinal movements of the plate.

A plate of uniform thickness, with different elastic moduli in the x and y directions, and which is squashed at the boundaries of x and y with the pressures p_x and p_y, becomes distorted. First the cross-sectional forces have to be integrated over the thickness h of the plate:

$$\mathbf{a} = \int_h \sigma \, dz \tag{3.15}$$

The vector $\mathbf{a}$ consists of three components: an x component, a y component and an xy component, that is, the contractions in the x and y directions and the shearing, where the xy component and the yx component are identical: $\mathbf{a} = \{a^x, a^y, a^x y\}$.

Conditions for equilibrium can be formulated from these expressions. If the plate is squashed by an external force exerting a pressure $\mathbf{p} = \{p_x, p_y\}$, this is balanced out by the internal cross-sectional forces. So we have

$$a^x_{,x} + a^{xy}_{,y} = -p_x \tag{3.16}$$

and

$$a^y_{,y} + a^{xy}_{,x} = -p_y \tag{3.17}$$

Quite apart from the integrals over the surfaces, action and reaction cancel each other out here.

To summarize this system of equations we will now use a system of matrix representation:

$$\begin{bmatrix} ,x & 0 & ,y \\ 0 & ,y & ,x \end{bmatrix} \begin{Bmatrix} a^x \\ a^y \\ a^{xy} \end{Bmatrix} + \begin{Bmatrix} p_x \\ p_y \end{Bmatrix} = \begin{Bmatrix} 0 \\ 0 \end{Bmatrix} \tag{3.18}$$

and thus simply

$$\boldsymbol{D\,\sigma} + \mathbf{p} = 0 \tag{3.19}$$

These cross-sectional tensions are brought about by external pressure, and lead to the distortions ϵ. The following equations apply:

$$\epsilon_x = \frac{1}{E_x h}(a^x - \nu\, a^y) \tag{3.20}$$

$$\epsilon_y = \frac{1}{E_y h}(a^y - \nu\, a^x) \tag{3.21}$$

$$\epsilon_{xy} = \frac{2(1+\nu)}{E_{xy} h} a^{xy} \tag{3.22}$$

This is Hookes Law. It comes from the definition of the elastic modulus E, which is defined as the proportionality factor between relative change in length $\epsilon = \frac{\triangle \mathbf{u}}{\mathbf{u}}$ and the cross-sectional force, that is force per unit area $a = F/A$:

$$\frac{F}{A} = E\frac{\triangle \mathbf{u}}{\mathbf{u}} \tag{3.23}$$

However, the shear deformation has been added, whose factor is the Poisson ratio ν. If the plate is compressed in the x direction, it also distends slightly in the y direction, i.e. it is "squeezed out." ν normally lies somewhere between 0.25 and 0.33.

These three equations of Hookes Law can also be written as a matrix.

$$\left\{ \begin{array}{c} a^x \\ a^y \\ a^{xy} \end{array} \right\} = \frac{h}{1-\nu^2} \left[\begin{array}{ccc} E_x & E_y\,\nu & 0 \\ E_x\,\nu & E_y & 0 \\ 0 & 0 & E_{xy}\frac{1-\nu}{2} \end{array} \right] \left\{ \begin{array}{c} \epsilon_x \\ \epsilon_y \\ \epsilon_{xy} \end{array} \right\} \tag{3.24}$$

This equation can be written in brief:

$$\boldsymbol{\sigma} = \mathbf{C}\,\boldsymbol{\epsilon} \tag{3.25}$$

The last item is the distortion, expressed as the relative change in length, in differential form:

$$\epsilon_x = \frac{\partial r}{\partial x} \tag{3.26}$$

$$\epsilon_y = \frac{\partial s}{\partial y} \tag{3.27}$$

$$\epsilon_{xy} = \frac{\partial r}{\partial y} + \frac{\partial s}{\partial x} \tag{3.28}$$

The result is, in matrix from:

$$\left\{ \begin{array}{c} \epsilon_x \\ \epsilon_y \\ \epsilon_{xy} \end{array} \right\} = \left[\begin{array}{cc} \frac{\partial}{\partial x} & 0 \\ 0 & \frac{\partial}{\partial y} \\ \frac{\partial}{\partial y} & \frac{\partial}{\partial x} \end{array} \right] \left\{ \begin{array}{c} r \\ s \end{array} \right\} \tag{3.29}$$

This has the short form:

$$\epsilon = D_\epsilon\, \mathbf{u} \tag{3.30}$$

If we combine these three equations, we obtain the basic equation for the plate:

$$\boxed{D\, C\, D_\epsilon \mathbf{u} = -\mathbf{p}} \tag{3.31}$$

Basically, we again have forces in equilibrium (or pressures in equilibrium), as described by this equation. The matrices D and D_ϵ consist solely of differential operators. D_ϵ is applied to $\mathbf{u}$. It is the relative change in length. C is the material matrix, i.e. the elastic modulus with the Poisson shear coupling, and thus contains only material data. Finally, D takes into account the fact that we have to differentiate over the whole plate.

This basic equation can be used as a basis for the FEM analysis. But it can also be the basic equation for the FDM. In the time-dependent procedure the cross-sectional force is calculated from the tension σ by using the displacement $\mathbf{u}$ and the external pressures $\mathbf{p}$ of the small "neighboring plates."

3.3 Discrete Mechanics

Discrete mechanics is different from continuum mechanics in that a continuum is only supposed for small elements. These elements are all linked to one another and thus constitute the whole body. This is necessary if the geometry of the body is so complicated that a general formula cannot be realized, or only with great difficulty.

Discrete mechanics has both advantages and disadvantages as compared to continuum mechanics, which shall only be outlined at this point, and examined in more detail in later sections. One disadvantage of discrete mechanics is that is has no general laws for the whole system. Once the geometry has been entered into the program, the latter calculates the desired results such as eigenvalues, strains, transients, etc. Whether the calculated results are based on laws applying specifically to the types of bodies examined, cannot be proved on the basis of the results, but can only be estimated. However, experience can compensate for this disadvantage. Here, as in all instrument acoustics, the reliability of the results and any systematic patterns in them can only be recognized in a comparison involving many calculations and a large amount of data.

This throws up a fundamental question relating to discrete mechanics: whether it is a theory or an experiment. Since the calculations are performed by a computer, it is not an experiment using an actual instrument, i.e. not a true experiment. But as the calculation is not subject to any metalaws, it has the much of the quality of an experiment. The programmed system must be checked for the influences to which it reacts, just like a genuine experiment.

This is the great advantage, at least of the Finite-Difference Method (FDM). It only assumes fundamental laws of mechanics applying locally, i.e. does not require any metalaws. It is precisely this property that makes the method so attractive. The whole system may possess surprising properties that are only recognized during the simulation. There are no limitations meaning that global laws are pre-supposed right at the beginning of the calculation. An example of such a global supposition would be assuming that the linkage of the string to the guitars top plate is a system with two mass points. The two-dimensional geometry of the top plate leads to transients, which are not covered by this simplification.

The idea is to simulate a complete system – in this case the guitar – on the computer, merely by determining forces, displacements, velocities, etc. *at every individual small element*, and so to have a vibrating system. The fundamental laws of mechanics can hardly be contested, and so the results achieved function as proofs, even if they provide results that are additional to the analytical calculations.

The result of this calculation is a sound. If this sound corresponds to that of a guitar, then the model is both correct and complete. If the sound does not resemble that of a physical guitar, if there are fundamental differences in their properties, or if the sound is incomplete, then we know which properties of the body are responsible for which (partial) aspects of the sound. This is the line of argumentation that is pursued in this book. It is precisely the absence of prerequisites, which is assumed here, that is part of the open view into the guitar and part of the proof offered of how it functions.

In fact, astounding results were achieved precisely in relation to this aspect. The method used has made some behavior very easy to grasp that would be difficult to interpret in an analytical way, because the movements of the instrument can be viewed in slow motion. For example, it often becomes clear why the sound happens to build up in the way it does. This is also of the greatest importance for the transient; its character is decisive not only for identifying the instrument as a guitar, but also for its musical expressivity. With the exception of vibrato during the stationary phase, the guitarist must achieve the desired quality of sound as the string is plucked. It is shown here that the attack of the fingers is also significant. The noise of the finger sliding over the string before letting go of it is very characteristic of guitars. The proportion of the sound attributable to this finger noise, and its form, contribute a great deal to the individual sound emanating from a guitar.

The advantages and disadvantages of discrete mechanics as opposed to continuum mechanics, FDM and FEM, the purely analytical solutions and the results, are discussed in depth later in this monograph. One first has to follow the theory of the FDM, as it is applied here, i.e. in a mathematical formulation.

4

Studies of the Guitar To Date

The guitar is generally described as a system of coupled vibrating systems. Both the strings and the body possess characteristic vibrating frequencies. The damping properties of the strings and the body force the guitar body to vibrate with the eigenfrequency of the string. The body is necessary to radiate the sound. The strings alone hardly radiate any sound at all, owing to what is called the "acoustic short circuit". Radiation of the sound always means that the air molecules are made to vibrate (Morse and Ingard 1983; Fletcher and Rossing 2000). If the string vibrates in one direction, the air particles in its way will be pushed aside. However, since a negative pressure is created on the other side of the string, the air particles are pulled more towards this region of negative pressure than they radiate the sound to the surroundings. So we cannot expect much radiation from the strings at all. The body is needed here to make the sound of the string audible.

This means that the guitar can be divided into problems concerning the vibration of the strings and problems concerning plucking, the body resonances, the coupling between strings and body, and finally of radiation.

In general stringed instruments – which include guitars – are located somewhere between wind instruments and percussion instruments. The vibration of the strings consists of a harmonic overtone spectrum with some inharmonic components. The harmonic spectrum corresponds to the vibration of the air, while the inharmonic components come from types of vibrations typical of percussion instruments, which are made of solid matter (i.e. membranes, rods, plates). The strings are also made of matter, but they are so thin and under such tension that they vibrate almost harmonically, which means that they can be approximated very well by a differential equation of the second order. The inharmonic components from thick strings containing lots of matter occur because of the strings rigidity, but they can be described by a differential equation of the fourth order, and can be ignored in the investigation of guitar strings. This harmonic spectrum must, however, be radiated by a vibrating system that actually has to be regarded as a percussion instrument with an inharmonic spectrum – the guitar body. This is the only part of the guitar that is large enough to avoid the acoustic short-circuit, because the stimulated air molecules cannot move either sideways or into the region of negative pressure underneath the top plate. Thus the only direction in which the air molecules can move is directly away from the top plate – that is, they radiate the sound away from the instrument (Morse and Ingard 1983).

4.1 Strings

Just like all strings, the strings of the guitar vibrate in several ways at once. The main type of vibration is transverse vibration. The rigidity of the string can also be taken into consideration. In addition there are longitudinal vibrations in the string. The vibration of the string continually alters the tension in the string very slightly and this means that slight fluctuations in frequency may be expected. Furthermore, this nonlinear effect must give rise to a small additional harmonic overtone spectrum. The same is true of the longitudinal vibrations. Since the string vibrates in three dimensions, resulting in two transverse and one longitudinal types of vibration, the bridge and the nut have to be described as systems with six degrees of freedom (David 1999).

4.1.1 Transverse Movement

The differential equation of the string is derived from its curvature. If our string has an initial tension T and we regard a small element of the string $\triangle x$ at a point x, this element is curved at an angle $\varphi(x)$ at its left boundary and $\varphi(x + \triangle x)$ at its right boundary. The backdriving force at this position is such that the left and right angles are different, i.e. the slope of the element varies along the string. Since a slope is the first spatial derivative, the backdriving force is given by the second derivative, that is the curvature. The backdriving force is now

$$F_R = T \sin(\varphi(x + \triangle x) - T \sin(\varphi(x)) \, , \tag{4.1}$$

or, since the rate of change of the slope of the string $_{,x}$ is

$$F_R = (T \sin(\varphi(x)) + (T \sin(\varphi(x)))_{,x} \, \triangle x) - T \sin(\varphi(x))$$
$$= (T \sin(\varphi(x)))_{,x} \, \triangle x \tag{4.2}$$

For small angles we can substitute $\sin(\varphi)$ with $\tan(\varphi)$ and we find what we had already assumed above, namely the second derivative for the displacement y.

$$F_R = T \, y_{,,x} \triangle x \tag{4.3}$$

We now apply Newton's second law of motion, stating that the backdriving force is equal to the accelerating force, $F_R = F_B$. This gives

$$T \, y_{,,x} \triangle x = \nu \, \triangle x \, y_{,,t} \, , \tag{4.4}$$

where ν is the linear density. Therefore it holds

$$y_{,,t} = \frac{T}{\nu} \, y_{,,x} \, , \tag{4.5}$$

where $c = \sqrt{\frac{T}{\nu}}$ s the velocity of the wave with the general dAlembert solution

$$y = f_{\text{right}}(ct - x) + f_{\text{left}}(ct + x) \, , \tag{4.6}$$

and assuming sinusoidal motion

$$y = A\sin(\omega t - kx) + B\cos(\omega t - kx) + C\sin(\omega t + kx) + D\cos(\omega t + kx) \tag{4.7}$$

with the wave number $k = \frac{\omega}{c}$. For our string of length L fixed on both sides it holds that $y = 0$ at the ends $x = 0$ and $x = L$. This gives the simplifications $A = -C$ and $B = -D$ if, for every t the end $y = 0$ is to remain motionless, and $\sin(kL) = 0$ or $\omega L/c = n\pi$ when $y(L, t) = 0$. The latter condition gives the eigenvalues $\omega_i = i\pi c/L$. So the transverse vibration of the string is described by

$$y_i(x, t) = (A_i \sin(\omega_i t) + B_i \cos(\omega_i t)) \sin\left(\frac{\omega_i \, x}{c}\right) \tag{4.8}$$

The last term is the space dependency of the string, with its eigenvalues ω_i. The first term is the time dependency, written using sine and cosine terms; alternatively, this can be expressed in terms of phases or the complex representation.

$$\begin{aligned} y_i(x, t) &= (A_i \sin(\omega_i t) + B_i \cos(\omega_i t)) \sin\left(\frac{\omega_i \, x}{c}\right) \\ &= (\tilde{A}_i \sin(\omega_i t + \varphi)) \sin\left(\frac{\omega_i \, x}{c}\right) \\ &= \tilde{A} \, e^{i\omega_i t} \sin\left(\frac{\omega_i \, x}{c}\right) \end{aligned} \tag{4.9}$$

In the second equation here we have the phase length of the time dependency, which was previously given by the values of A and B, now expressed in terms of φ, with the new amplitude $\tilde{A} = \sqrt{A^2 + B^2}$. In the third equation the complex notation is used, with $i = \sqrt{-1}$, corresponding to the first equation, which can also be viewed as a movement within the phase space. If in the first equation A corresponds to the vertical axis and B to the horizontal axis, in the third equation this would correspond to the division of the phase space into a real axis and an imaginary one.

4.1.2 Longitudinal Movement

In principle the longitudinal vibration of the string is derived in the same way as the transverse vibration. The difference here is that it is not the change in the slope of the string that corresponds to the backdriving forces, but the change in the pressure fluctuations along the string. Here the elasticity of the string begins to play a role. We define the Youngs modulus thus:

$$F/A = E\frac{\triangle y}{x} \tag{4.10}$$

that is, the relative change in length $\frac{\triangle y}{x}$ of the string of cross-sectional area A subject to a force F, where y s not the transverse displacement, but the longitudinal shift of an element. Since the accelerating force is given by

$$F_B = \varrho A y_{,,t} \tag{4.11}$$

the resulting differential equation is

$$y_{,,t} = \frac{E}{\varrho}y_{,,x} \, , \tag{4.12}$$

where $c = \sqrt{\frac{E}{\varrho}}$ of the wave speed.

So, the equation for the longitudinal vibration is of the same type as that for the transverse one, i.e. a second order differential equation. Its solution differs only in its angular frequencies ω_i, and not in the fact that the longitudinal waves also possess their own harmonic overtone spectrum. However, the fundamental frequency of the longitudinal vibration is considerably higher than that of the transverse vibration. It is about eleven times as high. In addition, it stimulates the bridge to produce mostly longitudinal waves, which cannot be radiated by the guitars top plate. But these longitudinal waves become bending waves in the ribs that are clearly audible. Furthermore, a longitudinal wave can twist the bridge, which causes bending waves in the top plate. Of course, the same is also true for the nut. The longitudinal vibrations of the string do not make up the major part of the radiated sound, but they do contribute to the higher frequency regions. We must remember that the sounds of instruments are actually described by their details, here by the contribution of the longitudinal waves, which can be seen very clearly in the case of the piano (Bank 2003).

There are extensive theories about strings, including not only their transverse movement but also their rigidity and their torsional movement, in (Pitteroff and Woodhouse 1998a; Pitteroff and Woodhouse 1998b; Pitteroff and Woodhouse 1998c; Pickering 1985). Here the theory of the bowed string is also discussed. The results of the calculations presented in this book are based on a purely transverse displacement. This is done because, coupled with the complete guitar, this produces results that illustrate the fine structure of the sound very well, even without the other aspects mentioned above. The first task here must be to exhaust these results. It will be possible to refine them at any later time but this will only be of use once the simplification has been fully understood.

4.2 Strings and Corpus Impedance

The coupling of the string to the top plate is described in terms of impedance. The characteristic impedance Z_0 of a string at one end $x = 0$ and at the other an infinite distance away is described as force $F(t)$, which has to be applied to get the velocity v at this point of the string

$$Z_0 = \frac{F(t)}{v(0,t)} \tag{4.13}$$

Unlike in wind instruments, the impedance is a purely real entity without any imaginary components. So, in this theory there is no phase difference between the string displacement at the point $x = 0$ and actuation. If we assume that the vibration of the string $u(x,t) = A\ e^{i\omega t - kx}$ with A being maximum amplitude and k the wave number, and if the actuation $I(t) = I_0 e^{i\omega t}$, then the characteristic impedance is (Fletcher and Rossing 2000)

$$Z_0 = \frac{T}{c}, \tag{4.14}$$

where T is the tension in the string and c the wave velocity of the string. This implies the assumption that the transfer of the force to the bridge is dependent on the angle of the string over the bridge. On the other side of the bridge the string bends downwards at a certain angle, here $45°$. The size of this basic angle is only important for consideration of the basic load imposed on the top plate by the string in a first approximation. The transverse vibration of the string back and forth causes this angle to change in the order of magnitude of the amplitude of the string where it meets the bridge. This change also leads to a change in the basic load, which is in phase with the change in the string, as indicated above. So the force

$$F(t) = -T\sin(\varphi) = -\frac{dT^y}{dx}, \tag{4.15}$$

is proportional to the derivative of the transverse displacement y in the x direction.

For a string with a finite length the calculation is more complicated. The result is a dispersive impedance, which is dependent on the frequency.

$$Z_{\mathrm{Ende}} = \frac{-ikT}{\omega} = -jZ_0 \cot kL \tag{4.16}$$

Here L is the length of the string and k is again the wave number. Richardson (Richardson 2003) describes an admittance that is the reciprocal of the impedance, i.e. a type of receptivity, as viewed from the point x where the string is plucked as

$$Y(\omega) = \frac{\sin(k'x)}{Z_L(\omega)\sin(k'L) + iZ_0'\cos(k'L)}. \tag{4.17}$$

Here Z_0 is the characteristic impedance, Z_L is the impedance at the bridge, L is the length of the string and Z_0' and k' are loss versions of the characteristic impedance and the number of waves, which take account of the damping using the Q factor. These are $k' = k\sqrt{1 + i/Q}$ and $Z_0' = Z_0\sqrt{1 + i/Q}$. Richardson speaks of being able to simulate all the possible properties of the guitar using an inverse Fourier transformation of the frequency-dependent admittance equation.

Woodhouse (Woodhouse 2003b) points out an important aspect affecting the impedance/admittance of the string and the body at high frequencies. This is a well-known problem[1], albeit "...somewhat a mystery." (Woodhouse 2003b) S. 140. In Woodhouses model the admittance is observed in the frequency space and then converted to a time series using an inverse FFT calculation. The resulting time series serves as the basis for analyzing the damping behavior of the individual overtones. These results can then be compared with the damping as measured.

In Fig. 4.1 below the frequency-dependent damping is presented for the frequency synthesis model and for measurement using Woodhouses method (Woodhouse 2003b). Amongst the more improbable explanations of this finding Woodhouse includes changes of temperature and humidity during measurement, the clamping of the guitar for making the measurements, and the influence of the bridge itself. He asks whether maybe the string does move sideways, or whether longitudinal waves are responsible for the enriched overtone structure (Woodhouse 2004).

The Finite-Difference Method presented here for the guitar also had this problem. After many experiments, however, it could no longer be supposed that the type of coupling conventionally proposed should be the only description of coupling in the guitar. The movements of the guitar top plate with the impedance coupling found up until now were mainly characterized by the impulse chain, which transfers the vibration of the string onto the body. A second order impedance theory was postulated as a solution to this problem. The physical basis for this is so obvious that the impedance calculation of the first order on its own cannot be accepted as the solution. A detailed representation of this calculation and the results it produces are included in the chapter on the results of the coupling of the string onto the top plate.

4.3 Guitar Body Modes

The three lowest frequency modes of a classical guitar body have been recognized by Meyer as combination modes (see also (Bader 2002a)). The top plate, the back plate and the air space vibrate as one body. This produces the three

---

[1] M. Karjalainen, of the Technical University of Helsinki-Eespo, also mentioned problems at high frequencies in the waveguide models. (Karjalainen, personal communication).

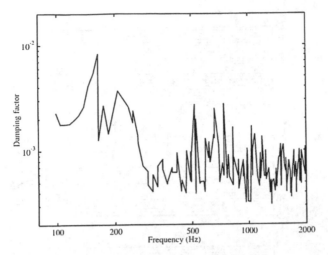

Figure 3a: *Decay rates for 6th string (E).*
Decay rates versus frequency, for all detected harmonics of
each note up to the 12th fret on the 6th string (E) of the test
guitar.

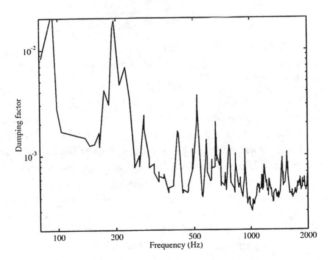

Fig. 4.1. Frequency-dependent damping values in Woodhouses frequency synthesis model (*bottom*) and the measurement (*top*) of the E string on the classical guitar. In the model the overtones show up too faintly in comparison with the measurement. Taken from: (Woodhouse 2003b) p. 139

lowest resonances (Meyer 1980). Meyer illustrates this with the figures reproduced below. Figure 4.2 shows the spectra of the top plate (top), the air space (bottom) and the coupled system of the top plate/air space (center). It is clear that the low resonance values of the top plate change considerably when it is coupled with the air space.

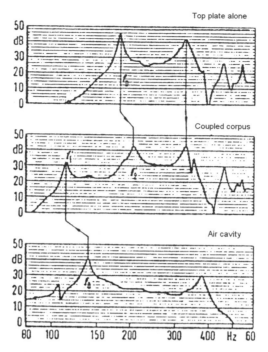

Fig. 4.2. Influence of the top plate and the air space inside the guitar on the body resonances according to Meyer (Meyer 1980). The mode s of the body model are combination modes

In addition to this, Woodhouse also discusses the influence on the body model of a body that is high and one that is low in relation to the top plate (Fig. 4.3), and finds as expected that the lower tuned back plate also leads to lower body resonances. However, its influence on the resonances as a whole is less than that of the top plate and the air space.

Fletcher & Rossing (Fletcher and Rossing 2000) considered the three lowest modes and also spoke of combination modes. In the case of the lowest mode, back plate and top plate vibrate completely out of phase with each other, and the air flows into the guitar as the parts of the instruments body vibrate outwards. In the case of the two higher modes, top plate and back plate are in phase with each other, and the air is sometimes in phase with the top plate and sometimes in phase with the back plate.

Numerous investigations of the frequency modes of the top plate have been carried out (Jansson 1971; Boullosa 1981; Christensen 1983; Christensen 1984; Rossing 1982; Richardson and Roberts 1985; Rossing 1985; Elejabarrieta et al. 2001; Derveaux 2002; Elejabarrieta et al. 2002). Naturally the modes of individual guitar elements, generally the top plate, back plate and internal air space, are different for every guitar. For this reason I shall present here

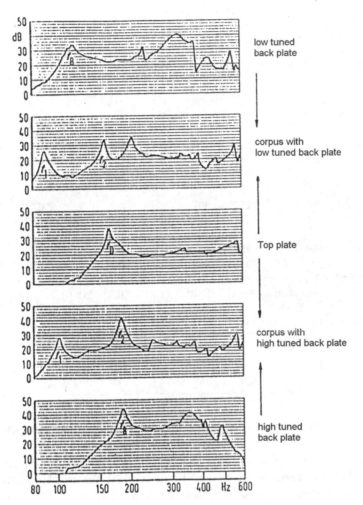

Fig. 4.3. Influence of the back plate on the corpus resonances. The spectrum of the top plate (*middle*) coupled with a back plate with high eigenfrequencies (*bottom*) results in a higher spectral peak of the coupled system (*second from bottom*), than in the case of a low tuned back plate (*top*), which results in a lowered peak of the coupled system (*second from top*). From: Meyer (Meyer 1980)

some examples of values for classical and western guitars in Table 4.1 (Rossing 1985).

Table 4.1 presents the individual modes in the form (K_l, K_q). K_l denotes the number of node lines that run along the length of the body and K_q denotes the number of the transverse node lines. So the mode $(0, 0)$ is the one without any node lines, in which the top plate (or the back plate) vibrates back and forth as a single unit. Details of the modes are discussed in the chapter on the

Table 4.1. Frequencies for the top plate, back plate, and the inclosed air for different classical and Western guitars. From: (Rossing 1985) S. 248

Top Plate	(0,0)	(0,1)	(1,0)	(0,2)	(1,1)	(0,3)	(2,0)	(1,2)
Folk								
Martin D-28	163	326	390	431	643	733	756	
Martin D-35	135	219	313	397	576	626	648	777
classical								
Kohno 30	183	388	296	466	558		616	660
Conrad	163	261	228	382	474		497	
Back plate	(0,0)	(0,1)	(0,2)	(1,0)	(0,3)	(1,1)	2,0)	(1,2)
Folk								
Martin D-28	165	257	337	369	480	509	678	693
Martin D-35	160	231	306	354	467	501	677	
classical								
Kohno 30	204	285	368	417	537	566	646	856
Conrad	229	277	344	495	481	573	830	611
Inclosed Air	A_0	A_1	A_2	A_3	A_4	A_5		
	Helmholtz	(0,1)	(1,0)	(1,1)	(0,2)	(2,0)		
Folk								
Martin D-28	121	383	504	652	722	956		
Martin D-35	118	392	512	666	730	975		
classical								
Kohno 30	118	396	560	674	780			
Conrad	127	391	558	711	772	1033		

results of the FDM calculation for plates and in the chapter on results of the steady-state of the initial transient.

Meyer quotes Letowski & Bartz (Letowski and Bartz 1971) concerning the harmonics of the three lowest modes. These two authors cite the harmonicity of the three lowest modes as the basis for a good guitar sound. But Meyer cannot confirm this in his research. However, it is common for guitar-makers and guitarists to speak about "A-guitars" or "E-guitars". These names do not mean "acoustic guitar" or "electric guitar", but instead indicate the basic frequencies of the three main resonances of the body in terms of pitch. According to various guitar-makers, by far the most common type of guitar is the A-guitar. The resonance frequencies of A-guitars are always a little lower in pitch than the A notes. The lowest resonance is slightly below the tone produced by the open A string. This is the note A, which is written for the guitar as a o b since music for the guitar is always written 8 octavo basso, means an octave higher than it is actually played.

The second resonance is a little below the a, for example the second fret on the G string, and the third slightly below a^1, for example below the fifth fret on the high e-string. This means that expressed in frequencies, the three resonances are slightly below 110 Hz, 220 Hz and 440 Hz. In the case of the second type, the E-guitars, these resonances would slip down below the corresponding E notes. Of course they are only approximate values, since it is not always possible to obtain the resonance frequencies for the exact corresponding values. But what is important is that they do not lie at the exact pitch of the fundamental of a tone that is played. If they did, there would be strong resonance with considerable magnification of the amplitude of the tone as compared to neighboring tones, and the corresponding frequency would dominate, i.e. it would sound much louder than the tones surrounding it. This makes the instrument almost impossible to play.

Therefore one important criterion for guitar quality, as for all instruments with a resonating body, is that the resonating behavior should be as smooth as possible, i.e. the amplification should be as even as possible over all the tones without so-called "dead spots". When speaking about guitars musicians call this balance in timbre. On the one hand this is an essential criterion for playability, as musicians especially in a concert situation have to trust on a smooth resonance behavior and strength of all note. On the other side, guitars which sound all the same over all playable tones are often described as cold, something characteristic is missing. This judgement holds also for other instruments like i.e. for the violin or other corpus instruments, where resonances occur as a chain of bandpass filters, which filter width are not necessarily equal. The timbre we are used with instruments never is exactly smooth. But just this characteristic deviations from smoothness makes the "color" of the instruments.

4.4 Nonlinearities and Transient Analyses

Despite of the guitar behavior mentioned above, in the guitar the fundamentals of the tones not adjacent to one of the fundamental resonances are also well pronounced. For this reason, the resonance curve of an instrument never exactly reflects the energy radiated by the instrument for the different frequencies, but is only an approximate guideline. (Woodhouse 2003b) Another important factor here is the precise type of coupling of the individual body elements. Nonlinearities can occur here that alter the resonant behavior to a considerable degree. Unfortunately the literature generally only deals with the linear case, what not seems to be the whole story. The related nonlinearities dealt with in this monograph occur mainly in the curved geometry and the coupling of longitudinal and transverse waves. It is worth considering whether a more far-reaching study of the guitar model presented here should be designed to include nonlinearities, for example in the actuation of the strings or the coupling of the string to the bridge. For an overview concerning nonlinearities in musical acoustics we first point to the literature (Fletcher and Rossing 2000), but a few important points have to be mentioned here, too. But at this point the reader is reminded that the guitar is also a system in the sense of synergetics (for information on synergetics, see (Haken 1990; Haken 1983)).

Here two linear vibration systems are regarded as one synergetic system if they are coupled in a nonlinear manner. As the system of equations starts with the displacements of the vibrating systems, the theory proposed here of impedance coupling of the second order is a nonlinear coupling of this type. For these couplings the rule is that the system with less damping forces its eigenvalues (here its frequencies) onto the system with more damping. In the case of the guitar it is the string that forces its frequencies onto the body. This is because the string is subject to much less damping (see the chapter on damping). Of course this phenomenon is also found in almost all instruments in which vibrating systems are coupled. The wind instruments (aerophones) are the most interesting in this respect. For example, the saxophone has a column of air that is subject to much less damping than the reed. So the column of air does not vibrate at the frequencies of the reed, but the reed is forced to vibrate with the values of the column of air. This synergetic relation is reversed in the case of organs with reed pipes. Here the column of air is much slower to react and therefore more severely damped than the reed of the pipe, which consists of a thin metal lamella. This is what makes it possible for the organist to tune the pipes by shortening or extending the vibrating length of the reeds using the tuning rod. The pitch of a note played on the saxophone is changed by altering the length of the vibrating column of air – by opening and closing the keys – and here the column of air is less well damped, which means that it determines the pitch of the note played.

However, most of the literature on nonlinearities in music is limited to the actuation; if we follow the reasoning of a musical instrument model consisting of a generator (i.e. a reed functioning as a ventil) and a resonator (i.e. a tube with an air column), the nonlinearities mostly appear in the generator. Nonlinearities are discussed in the case of the violin (Woodhouse 2003a; Pickering 1989; Schumacher and Woodhouse 1995a; Schumacher and Woodhouse 1995b). The actuation mechanism in the violin is realized by the interplay of the string and the bow. The bow is coated with a type of resin (colophony, also called rosin). This material is solid at room temperature, but liquefies at temperatures not much above normal ambient temperature. If the bow moves over the string, the string initially moves with the bow. This is caused by static friction. Once the string comes away from the bow, it slides back towards its original position and heats the rosin on the bow. During this period there is dynamic friction between the bow and the string. If the string and the bow come to rest relative to each other, then the string "sticks" to the bow as the rosin becomes hard again. The precise course of this sticking and sliding is a hysteresis loop, that is, a marked nonlinearity. The conditions required for the bow to "let go" of the string in the stationary case, with Helmholtz saw-tooth vibration, are determined by the differences in velocity and displacement between the string and the bow. This is the only way that the string can force its eigenfrequency onto the system of string+bow, through this nonlinear coupling. In this case the order parameter, that is, the value that determines from what point onward this "coercion" takes place, is the pressure of the bow. The saw-tooth only forms once a certain bow pressure has been exceeded. The period of vibration is halved at this pressure, i.e. bifurcation, called flageolet. Bow pressures that are much below the normal cause chaotic noise. On the other hand, if the bow pressure is increased greatly, the first effect is a doubling of the period and the suboctave of the saw-tooth tone, followed by the "coercion" as pressure increases further. Then the pitch of the note played is no longer dependent upon the length of the string, but is a direct function of the bow pressure. There are several studies of the violin which found that increased bowing pressure caused subharmonics, including (Cremer 1974; Guettler 1994; Hanson et al. 1994; Kimuar 1999).[2]

Such phenomena in the guitar can only be mentioned in passing in this monograph. However, on the one hand they come about through the complicated geometry of the guitar, and on the other through the couplings of longitudinal and transverse waves. Since the couplings of the parts of the guitar are realized by transferring a force, which in the case of the longitudinal wave is a derivative of the second order, and in the case of the transverse wave a derivative of the fourth order, all the couplings are to be regarded as nonlinear. An interesting case occurs if the damping of longitudinal and transverse movements is different. Then, for example, the back plate can actually be forced to vibrate with the longitudinal waves of the ribs. These two systems are linked by a 90° coupling. Therefore there has to be some conversion of the longitudinal movements in the ribs to transverse movements in the back plate. This way it would be possible to make the longitudinal, much higher frequencies from individual guitar elements audible in the sound, by "coercing" transverse parts, the only ones to radiate a significant proportion of the sound, to vibrate with the longitudinal movements. The second possibility for coupling transverse and longitudinal waves in the geometry occurs at two places in the guitar. First, the ribs are curved, and secondly, the back plate bulges outwards owing to longitudinal basic tension. If the results of the calculations for the curved geometries are compared with the simplified calculations for the non-curved geometries (see the chapter on curvatures), then not only can an increase in the eigenvalues of the geometries be determined, but the brightness of the sound is clearly enhanced. This can be regarded as a type of coercion in the synergetic sense. The fact that this is not complete here is due to the complexity of the geometries. In the example of the column of air inside a saxophone, which is coupled to the reed, the coupling occurs at one place. The rest of the column of air vibrates on its own. And yet it is still possible to observe that the reed does not only vibrate with the modes of the column of air, but – primarily at its tip – it still vibrates at its own

---

[2] Sample program demonstrating the coupling of the bow and the string can be downloaded from *http://www.suul.org/download*. It was created by the author in C++ for Windows and made available to students at the university for teaching purposes. The bow pressure can be varied and the resulting vibration of the string and the saw-tooth movement can be observed at the bridge.

frequencies. The coupling of the two possible types of vibration in the guitar back plate or the ribs takes place all over the back plate and the ribs. Coercion can occur alongside the normal vibration.

During further investigations of the string, Hanson et al. found periods of some seconds, which they explained in terms of nonlinearities (Hanson et al. 2003). An explanation of the precise causes is not given here, but the experiments reveal the phenomenon clearly. The extent to which this also occurs in an iteration calculation such as ours must remain undetermined initially, although it is possible to ascertain. The experiments show the string under tension. To what extent this tension makes such long periods possible, cannot be determined here. However, this would have to be taken into consideration in a simulation using FDM. In this book the phenomenon was not found.

Another occurrence of nonlinearities in musical acoustics is found in strong actuation of instruments. With wind instruments the sound is created by artificial blowing in excess of the normal value (Olivier and Dalmont 2003; Voessing et al. 1993). his causes bifurcation spectra like those also used by instrumentalists as multiphonics produced by overblowing. This enables the player to play several notes simultaneously.[3] He same phenomenon is also found in percussion instruments (Touze and Chaigne 2000; Fletcher 1985; Rossing and Fletcher 1983; Gibiat and Castellengo 2000). Here, strong actuation creates a bifurcation spectrum. In addition, these severe displacements change the pitch of the instruments. In contrast to linear displacement, where Hookes law still applies and thus a sinusoidal vibration arises, in the case of nonlinear displacement the sine wave is deformed. This is because the greater the displacement, the more disproportionately large the backdriving force. The result is a back driving that occurs faster than in the linear case, i.e. the period is shortened. But this shortening must die away as the displacement diminishes. This leads to a pitch-glide, which usually goes from higher frequencies to lower ones. One example of this would be the Chinese gong at the Peking Opera. It is struck several times in rapid succession in order to create dramatic effects. As the metal of the gong is very thin, when the displacement is great it begins to behave in a nonlinear fashion. The pitch-glide sound thus serves to create noticeable effects because of the repetitions of the tone. The same effect can occur with severely displaced strings, such as in the harpsichord, whose strings are pushed a long way up and then be suddenly released at the beginning of the tone (Beurmann and Schneider 2003). Here the nonlinearity leads to a greater fullness of the sound. This is because the deformed sinusoidal vibration described above contains an overtone spectrum precisely because of its deformation, and this overtone spectrum is harmonic owing to the continuing periodicity of the sound. Such nonlinearities in the

---

[3] For a systematic view of possible multiphonics in the clarinet (a) by overblowing, (b) by underblowing and (c) by combining keys, see Krassnitzer (Krassnitzer 2002) who includes a collection of 848 possible multiphonics with up to seven tones played at once.

guitar have not yet been included in the calculations in the model presented here. The reason for this lies in the large number of possible types of vibration of the system. This case examined here has already demonstrated such a high degree of complexity that these phenomena can only be included in later studies.

One further possibility for nonlinearities is the geometric one. The Chinese gong, just like cymbals, crashes and certain sound bowls, are made by hammering, i.e. its surface is marked with many small impressions. These impressions lead to the energy being transferred from the deep modes to the high ones. This means that after the instrument has been struck, in the steady-state there is a marked increase in the brightness of the inharmonic overtones, which results in the typical sound of a large gong. This geometric coupling corresponds to the curvatures of the ribs and the back plate of the guitar, and also to the 90°-couplings of the top plate to the ribs and of the ribs to the back plate. The phenomena occurring here are likewise discussed in the chapter on curvatures.

But also other geometric properties of sounding bodies can lead to nonlinearities, such as in kargyraa, Tuvan undertone singing, which uses the same principle as blues undertone singing (Gibiat and Castellengo 2000; Bader 2002a). The tones are created by distorting the vocal chords. (On the singing voice in general, see (Berry et al. 1996; Lee et al. 1998; Neubauer et al. 2004; Fitch et al. 2002; Wilden et al. 1998)). The same phenomenon can be seen in multiphonic tones of wind instruments (Backus 1978; Krassnitzer 2002), where the openings for the keys disrupt the flow of air. These disruptions also lead to bifurcations, which means that the player can play several tones at once. But there are also instruments where this disruption is an essential part of their body. One example would be the Chinese Dizi flute (Tsai 2003). Its top hole is covered by a membrane, which does not lie quite tight against the flute, but instead has some freedom of movement. The result is a sound that has two, four or even six overtone spectra surrounding the middle frequency of the pipe, and the distance from these overtone spectra to the fundamental frequency is $f(n, \pm) = f_0 \pm n * f_B$. Here f_B is the difference between the frequencies of individual fundamentals of the overtone spectra. This difference is the same for all fundamentals. These differences lie above and below the fundamental frequency f_0 the pipe; there are n of them, where n is between two and four, depending on the pressure of attack. Such types of nonlinearities do not occur in the guitar.

One additional aspect of the theoretical discussion of nonlinearities can be found in Fletcher (Fletcher 1978) in connection with the phenomenon of mode-coupling. This is an attempt at a mathematical explanation of the observation that the overtones produced by musical instruments are not in a rough harmonic relationship, but in a very exact one. For some instruments this is difficult to explain, for example in the case of brass instruments. The conical body of these instruments spreads the overtones out, away from one another. The bell at the end of the instrument compensates for this by bringing the overtones closer together again. One would not expect these phenomena to cancel each other out exactly. Fletcher regards the reason for the overtones and the modes being in exact harmonic relationships to one another, that is, for mode locking to occur, as a large nonlinear actuating force. From this he develops a theoretical model that can be applied to several families of instruments. In the case of wind instruments we can imagine the steady-state vibration to be such that above a certain blowing pressure, the pressure impulse emanating from the mouthpiece through the pipe is pulled apart by the dispersion, and is "forced back together" again into an impulse when it is reflected at the mouthpiece. Interestingly, this phenomenon also occurs in the guitar string we are investigating. Once the impedance coupling of the first order can no longer be recognized as serving the purpose, the components of the spatial wave in the string become responsible for the overtone structure of the string, and these parts mask the fundamental vibration with their damping and reflection (see the chapter on results from the coupling between string and bridge). However, the fundamental period forces these components of the spatial wave to adopt its own periodicity. This results in a far richer overtone spectrum, which is strictly harmonic in relation to the fundamental frequency. This is a very important finding of this book and its significance for later research should not be underestimated.

Nonlinearities are highly significant in musical acoustics and they occur in abundance. The component of the sound that is most affected by these phenomena is the initial transient. Here there are many nonlinear backdriving forces. In addition, a quasi-stationary component has not yet formed, so there may be the most varied phenomena, highly specific to the particular instrument. A summary of works about initial transients can be found int (Reuter 1995). Especially for the violin, see (Gueth 1980; Gueth 1995; Roberts 1986; Rodgers 1991). Visualizations of transient processes – unfortunately again only by means of artificial actuation – are provided by Molin et al. (Molin et al. 1990; Molin et al. 1991). Using interferometry, the authors show that the violin is a dipole radiator for approximately the first 30 ms nd only then does it become a monopole radiator. The violin is viewed as a monopole radiator since its sound post hinders vibration of the left half of the instrument. So most of the sound is radiated from the right half of the violin, resulting in the sum total of the energy radiated being considerably higher than it would be for a dipole radiator, where total dissipation occurs. Furthermore, Molin finds that the violins f-holes greatly influence the spread of the waves, as expected, since the approaching wave breaks at the holes. Of course this is also observed in the guitar. The first wave front of the initial transient breaks up at the sound hole and other geometric discontinuities. The research into the specific modes of behavior of musical instruments during their initial transients are many and varied. On the subject of chaoticity and nonlinear behavior of musical instruments, Bader (Bader 2002a) found fractal dimensions to give a measure of the chaoticity of the initial sounds in guitars, violins, wind instruments and percussion instruments. The author mainly uses fractal correlation dimensions, but also some information dimensions and information structures. It is possible to use the tone register to ascribe the sound character of the instruments to musical expression and to families of instruments in relation to the size and distribution of the fractal dimension.

4.5 Discrete Instrument Models

Finite-Element Analysis in musical acoustics was first applied to the violin. (Roberts 1986; Knott 1987; Rodgers 1991; Bretos et al. 1999; Tinnsten and Carlsson 2002).

The guitar has been investigated by Elejabarrieta (Elejabarrieta et al. 2001; Elejabarrieta et al. 2002) und (Derveaux 2002). Elejabarrieta et al. studied the static eigenvalues of the guitar. They initially turned their attention to the construction of the guitars top plate in its various stages, and accompanied this with FEM analyses. They were able to determine changes in the eigenvalues as the top plate was progressively pared down and following the addition of bracing. The second publication contained a calculation of the eigenvalues of the whole guitar. The three lower combination modes were determined.

5

The Discrete Guitar Model

The modelling of the classical guitar, as suggested in this work, consists of the coupled parts of the guitar. Table 5.1 shows the six guitar parts and the corresponding coupled parts: top plate, back plate, ribs, neck, enclosed air and strings.

Table 5.1. The coupling between the different guitar parts

Guitar Part	Coupled with
top plate	ribs
	enclosed air
	neck
	strings
back plate	ribs
	enclosed air
ribs	top plate
	back plate
	inclosed air
enclosed air	top plate
	back plate
	ribs
strings	top plate
	neck
neck	top plate
	strings

All guitar parts show different material properties. These properties lead to arbitrary differential equations, which hold for the several guitar components. There are two fundamentally different differential equations, one for the bending and one for the longitudinal or in-plane movement. The equation for the bending is a differential equation of fourth order, the longitudinal equation is of second order both with respect to space. As both are wave equations, both are of second order with respect to time. The description of the mathematical equations and their discrete versions is treated further down in the discussion. At this place, we shall first give an overview of the reasonings for using which equation for which guitar part (see Fig. 5.2).

5.1 Top and Back Plate

The top plate is treated as a plate. In plates, the reason for a backdriving force is the bending in its orthogonal direction, which is described by a differential equation of fourth order. But a plate also knows in-plane or – now and in the further discussion called – longitudinal waves. These waves are of second order.

These two movements are coupled, as the bending wave causes a structural change in the plate, which has impact on the longitudinal movement. Vice versa, the same influence holds, too. The strain of the longitudinal waves feed back into the bending waves. So both differential equations are implemented as coupled right from the start and not each one for itself with i.e. just an additional coupling force term.

5.1.1 Coupling within Top and Back Plate

The bending waves of the top plate, the back plate, the ribs or the neck are the only movements, which lead to a radiation of energy from the guitar body, if we neglect the small areas around the plates (top plate, back plate, ribs) which represent the hight of the plate. So the longitudinal waves have nearly no influence of the radiated sound, but are considered here out of two reasons.

First, as the two fundamental movements are coupled, there is an indirect influence of the longitudinal waves to the radiated sound via the coupling to the bending waves, which form the guitar sound. So the bending wave movement is not undisturbed, which leads to a change in the details of the radiated sound. We shall remind here, that the main aim of this study is the fine structure of the guitar sounds, as these are considered as a measurement of quality between different guitars by both guitar players and makers.

The second reason for the inclusion of the longitudinal waves here is the coupling between top plate and back plate via the ribs. This coupling is looked at as the 90° angel between top plate (back plate) normal vector and ribs normal vector at the boundaries of the ribs causes a transformation of the longitudinal waves of the top and back plates respectively in bending waves of the ribs. And this angle also transforms the longitudinal waves of the ribs into bending waves of the top and back plate. Figure 5.1 shows an example. Here, a bending wave of the rib is transformed into a longitudinal wave of the top plate (or vice versa).

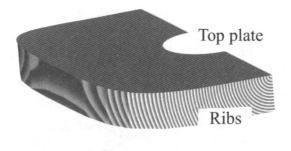

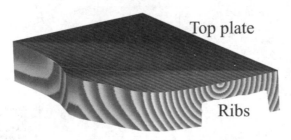

Fig. 5.1. Schematic representation of an example of a coupling between two guitar parts, here the guitar top plate and the ribs. **(a)** guitar body without displacement of any guitar part in relation to **(b)** the top plate is displaced in a longitudinal way in its plane, which leads to a bending displacement of the guitar ribs

5.1.2 Boundary Conditions of Top and Back Plate

The boundary conditions of the guitar are described in the literature as a mixture between fixed and clammed boundaries. Indeed, the top and back plate are fixed at the ribs due to gluing, but on the other side, these boundaries can move to a certain extend, so a force-coupling to the ribs can take place. These findings are represented in this model by letting the elements of the plates which are coupled to the ribs move freely, but fixing two elements, which are further out of the geometry, behind the ribs. So the top plate is fixed, but can couple at all time to the ribs in all bending and longitudinal movements. These boundary conditions are also applied to the ribs.

5.1.3 Mass Distribution of the Top and Back Plate

The top and back plates are considered of being of constant thickness. Although a complete mass matrix is at work in the implementation, with with at each point of the plate any small change of thickness can be modelled, it would be useful for the reconstruction of a guitar, to have a top plate which is cut and prepared and judged by a guitar builder as being perfect balanced

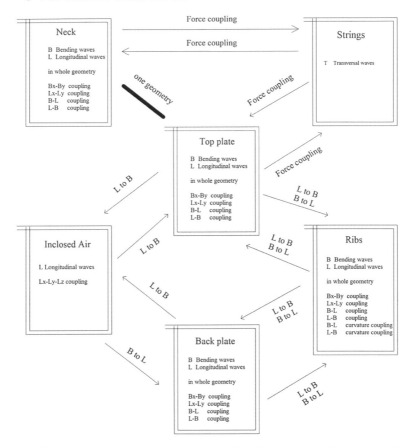

Fig. 5.2. Schematic figure of the guitar model. The single guitar parts, like top plate, back plate, ribs, neck, strings and air inclosed in the guitar are shown in their coupling relations. In the boxes, the possible ways of moving of the guitar parts, the internal possibilities of vibration and the couplings of the differential equations of the different guitar parts are shown; **B**: bending movements, always normal to the plates geometry, is radiated; **L**: longitudinal or in-plane movements (except of the air inclosed in the guitar, where **L** is the only possible wave, because air do not have a shear modulus), is not radiated; **T**: transversal movement of the strings; the indices "x", "y" and "z" are spacial coordinates, i.e. **Bx**: bending movement in x-direction. The top plate is moved by the strings in a force/anti-force coupling. The top plate vibrates in bending and longitudinal movements and so attracts the other guitar parts to their movements (for a closer explanation see text)

in sound quality. This step in not realized in this work although it should be done in the future.

But the mass matrix is still in use here, because it is needed for the bars and fan bracing of the top and back plate. Here, the additional masses at the different places of the top plate are used to model these elements.

5.2 Ribs

The ribs are curved in shape, where they differ from the other used plates, which are flat. On the other side, the rib is still a plate, like the top or back plate. So here, we can use the same reasoning which has been used above with the top and back plate. Additionally, the curvature of the ribs have to be taken into consideration.

As a curvature can be described as a differential equation of second order with respect to space, the fundamental mode of vibration of the ribs is still a bending wave with a differential equation of fourth order, the resulting bending-differential equation of the ribs is of sixth order along the spacial direction of the curvature. This differential equation of sixth order uses the fact, that at each point of the ribs there is a curvature of the geometry, which can be added in the used equation.

Furthermore, this curvature leads to another coupling, which is the one between longitudinal and bending waves. As has been shown above with the coupling between top and back plate with the ribs, the 90° − "curvature" leads to a complete transformation of bending to longitudinal movements, which is easily verified experimentally. An angle of curvature, which is less than 90° as is in place with the ribs curvature, leads to a partial transformation of this kind, depending on the profile of the curvature at place. I.e. this additional coupling based on curvature leads to a steady energy flow between the longitudinal movements of the ribs and the radiating bending waves, which takes place in addition to the coupling between longitudinal and bending movements existing in the ribs because of the ribs being a "plate" as discussed above. The longitudinal movements of the ribs can be caused by the top or back plate moving in bending ways.

The ribs are looked at as a closed, kind-like eight-shaped structure of the guitar body, where the masses of the blocks at the guitar body's end and at the connection of the body to the neck is implemented in the mass matrix as additional terms.

5.3 The Neck of the Guitar

The neck of the guitar is a bending stiff structure like the top or back plate or the ribs of the guitar. So the ruling differential equations and the internal coupling between longitudinal and bending movements are the same as in the mentioned other guitar parts. The coupling of the neck to the guitar body is also implemented as a coupling between bending waves and longitudinal movements. The neck is taken as a prolonged top plate. The coupling between neck and strings works as a force impact. The boundary conditions of the neck at its sides and at the necks head are free.

5.4 The Strings of the Guitar

The strings of the guitar are taken as a differential equation of second order. Although the bending stiffness and the torsional modes of vibration take place in the string, too, for a first implementation of the guitar they are neglected here. The reason for this is the complexity of the guitar behaviour and its sound production even without theses additional terms, which later then could be added. As the complexity of couplings of the different guitar parts lead to a lot of complex behaivour, we are first interested in this complexity. To be sure, it is not a result of the bending or torsional behaviour of the string, we neglect the bending and torsional string behaviour here at first.

The coupling of the strings at the top plate and the neck is a force impact. Indeed it showed up, that a simple angle relationship of string and bridge is not enough to simulate the force impedance correctly. We need a second order term here, which actually comes out of the wave equation itself, for to transmit the higher frequencies correctly. This force interaction of string and guitar of second order is to be demanded out of physical reasons of simple geometry discussion (for details see text below).

5.5 Air in the Guitar

The air in the guitar is described by a differential equation of second order completely, because air do not have a Poisson contraction and therefore no bending stiffness. It is the only guitar part in the present work, which is looked at as a three dimensional space. Thereby, all three dimensions are implemented as coupled, like in the two-dimensional cases of the plates of the top and back plate, the ribs or the neck. The air therefore is a room which in the FD-model is made discrete through node points, which are in the same distance to one another. Its boundary conditions at the contact areas with the top and back plate and the ribs are modelled in a way like the boundary conditions of these mentioned parts. The node points which couple to the other parts are free, the further (virtual) points over the boundary are fixed to represent the special boundary conditions of a fixed end, which still can act and react to forces and move to a certain extend. So the coupling of the air in the guitar to the top and back plate and the ribs is thought of as a force interaction. The only exception here is the sound hole with a free boundary condition.

5.6 Sound Radiation of the Guitar Body

The sound radiation of the guitar body is calculated as an integral over each single guitar part, where a weight is used for the point in the room, where the sound is looked at (maybe a microphone). Here, just bending velocities are used for the calculation, because only moving node points of the bending movement are able to act on the air around the guitar.

The weight used here of the different guitar parts depends on the damping of the air, which with further distance from the guitar leads to a reduced sound pressure intensity. In many trail experiments it could be found, that this modelling leads to the correct results; although the sound intensity differences with small microphone distance to the instrument are just small, they have to be taken into consideration here. Otherwise the integrated sound is nearly not present, because of the extinguishment between the mode shapes of positive and negative phase.

The weight of the different parts of the guitar to the integrated complete sound depends on the room. Here the following values are used:

guitar part	dB in relation to the air inside the guitar
air inside guitar	0 dB
top plate	−3.2 dB
back plate	−6.5 dB
ribs	−8.5 dB
neck	−12 dB

The sound energy out of the air room inside the guitar radiated through the sound hole is the highest of all guitar parts. Therefore with guitar recordings the fraction of the sound from the air inside the guitar compared to the sound from the other guitar parts is mostly quite prominent and so suited as a reference for the top and back plate, the ribs and the neck. As it is assumed here, that the microphone position is in front of the guitar (in a distance of $d = 1\,\text{m}$ in front of the sound hole), the top plate is the second important part. As shown in the results chapter, the top plates radiated spectral centroid is much higher than that of the radiation of the sound hole. So although the sound pressure level of the top plate is lower than that of the sound hole radiation, the top plate is heard more prominent, because of the higher psychoacoustic sensibility of the ear in the main frequency region of the top plate. The back plate is next, because of its big radiation area, followed by the ribs and the neck. The logarithmic dB unit is used here, as the ear also perceives in a logarithmic scale. So the relations of the sound pressure levels are approximated by a psychoacoustic reasoning, except for the top plate, out of reasons mentioned above, but also because its attack is very fast and so perceived much louder than the other parts, too. As shown in the results section, the top plate is therefore responsible for the sound attack; it is perceived much more prominent as the other guitar parts, which are responsible for the "tone color" (speaking in musical terms) and the radiation from the sound hole, which adds a "pressure" or "pumping" to the sound.

For a description of the differential equations we shall now begin with the most simple case of the string. The string is thought of being one-dimensional and is treated as a differential equation of second order. So the string is the easiest geometry to differentiate.

5.7 The String Movement and Its Discrete Form

The string is divided into N node points. For every node point $P(n)$ with $1 < n < N$ the displacement $s(n)$ has to be calculated for every point in time $t(m)$ with $0 < m < M$, if M is the overall amount of time steps to be calculated. The differential equation of the string is

$$\frac{\partial^2 y}{\partial t^2} = a = \frac{T}{\mu}\frac{\partial^2 y}{\partial x^2} = k_S\frac{\partial^2 y}{\partial x^2} \ , \tag{5.1}$$

with a being the acceleration (here mass-weighted) and k_S as the material constant of the string, also

$$k_S = \frac{T}{\mu} = c^2 \tag{5.2}$$

with the wave speed c of the string.

These differential equation is written discrete in the form

$$a = k_S\frac{s(n-1) - 2s(n) + s(n+1)}{\Delta x^2} \ . \tag{5.3}$$

This differential fraction is the discrete form of the differential equation. If we substitute the infinitesimal variable dy by the discrete variable Δy, then dx also becomes discrete in the form of Δx, because of

$$\frac{dy}{dx} = \lim_{x\to 0}\frac{\Delta y}{\Delta x} \ . \tag{5.4}$$

Now Δy can also be written as a difference of two displacements $s(n)$ and $s(n+1)$ at the points $x = n\Delta x$ and $x = (n+1)\Delta x$.

$$\frac{\Delta y}{\Delta x} = \frac{s(n+1) - s(n)}{\Delta x} \ . \tag{5.5}$$

The second derivation follows in the same way like

$$\frac{\Delta^2 y}{\Delta x^2} = \frac{\frac{s(n-1)-s(n)}{\Delta x} - \frac{s(n)-s(n+1)}{\Delta x}}{\Delta x} \tag{5.6}$$

$$= \frac{s(n-1) - 2s(n) + s(n+1)}{\Delta x^2} \ . \tag{5.7}$$

Here it is assumed, that around the node point n with the displacement $s(n)$, there are two node points $n-1$ and $n+1$. Here, two slopes appear at the left and the right side of the middle node point n, which have the displacement differences $s(n-1) - s(n)$ and $s(n) - s(n+1)$ and the step Δx

in the x-direction. The difference of these slopes – again along Δx – leads to the second derivation.

These differential equation holds for all elements $n = 1$ to $n = N - 1$. For the elements $n = 0$ and $n = N$ additionally the boundary conditions have to be applied for the fixed string like

$$s(0) = 0 \quad \text{and} \quad s(N) = 0 \quad \text{for all } t \tag{5.8}$$

for every point in time t.

If we know the (mass-weighted) accelerations a, because of $acceleration = velocity/time$ of $a = \frac{v}{t}$ it is

$$v(n) + = a \, \Delta t \tag{5.9}$$

with Δt as the applied time interval. This is an iterative equation. So to a velocity $v(n)$ of the previous point in time t, the change of velocity $a\Delta t$ is added. So for a velocity v_{t+} at the following point in time t_+ we can also write

$$v_{t_+}(n) = v_t(n) + a \, \Delta t \ . \tag{5.10}$$

The change of displacement now is because of $velocity = lenghth/time$ or $v = \frac{s}{t}$

$$s_{t_+}(n) = s_t(n) + v_{t+}(n)\Delta t \tag{5.11}$$

or

$$s_{t_+}(n) = s_t(n) + (v_t(n)\Delta t + a\Delta t^2) \ . \tag{5.12}$$

So the fundamental method of discretion is described now. The last step from the acceleration to the calculation of the velocity and the displacement are all the same with all forthcoming discrete equations. So the task is, for complicated geometries to calculate the accelerations. In the following examples we will do this for the geometries present in the guitar.

5.8 Longitudinal Waves and Their Discrete Form

The longitudinal movement in the plates plane can be looked at as a differential equation of second order. This equation has the form of the differential equation of the transverse string displacement

$$\frac{\partial^2 y}{\partial x^2} k_L = \frac{\partial^2 y}{\partial t^2} \tag{5.13}$$

with the material constant

$$k_L = \sqrt{\frac{E}{\varrho}} \ . \tag{5.14}$$

The discrete form follows therefore as has been shown above with the string.

But in the case of the plate, additionally we have to consider shearing and Poisson contraction. These are written in discrete form as displacement differences of the node points around the middle node for which the calculation takes place. The mass-weighted acceleration

$$\mathbf{a} = \{a^x, a^y\}, \tag{5.15}$$

which we have to calculate in a way described in the above section, now splits into three parts

$\mathbf{a_L}$: acceleration caused by the displacement differences just in the respective longitudinal direction
$\mathbf{a_S}$: acceleration caused by shearing
$\mathbf{a_Q}$: acceleration caused by Poisson contraction

for each of the two room directions x and y like

$$a^x = a_L^x + a_S^x + a_Q^x \tag{5.16}$$

$$a^y = a_L^y + a_S^y + a_Q^y . \tag{5.17}$$

Here $\mathbf{a_L} = \{a_L^x, a_L^y\}$ is the acceleration caused by pure contraction in the respective direction x and y without coupling between these directions, $\mathbf{a_S} = \{a_S^x, a_S^y\}$ is the shearing and $\mathbf{a_Q} = \{a_Q^x, a_Q^y\}$ is the Poisson contraction.

The shearing can be described through the displacement differences in the node points perpendicular to the respective node point *in direction* of the respective displacement of the node point to be calculated. For example, if we look at the node point P_0 with the coordinates x_P and y_P in the x- and y-direction and the displacements in the x-direction P_x for the next point in time has to be calculated, then for the shearing in x-direction, the node points P_{y-1} and P_{y+1} have to be taken into consideration. This is shown in Fig. 5.3.

If i.e. two node points P_{y+1} and P_{y-1} have a positive x-displacement, then they tear the node point P_0 in the positive x-direction, if the x-displacement of this point is less then that of the two node points.

$$a_S^x = \frac{1}{2}(1 - \nu)\frac{x_{P_{y-1}} - 2x_{P_0} + x_{P_{y+1}}}{\Delta y^2} \tag{5.18}$$

and

$$a_S^y = \frac{1}{2}(1 - \nu)\frac{y_{P_{x-1}} - 2y_{P_0} + y_{P_{x+1}}}{\Delta x^2} \tag{5.19}$$

with the Poisson number ν, which is given as a relation between the Young's modulus E and the shear modulus

$$G = \frac{E}{2}(1 - \nu) . \tag{5.20}$$

The Poisson contraction is calculated from the strain, which acts on the node point in question, i.e. as shown in Fig. 5.4 in its y-direction, which

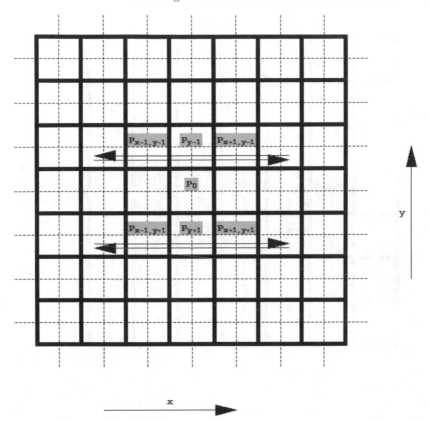

Fig. 5.3. The influence of the shearing around the node point P_0 on the change of displacement of this node point is described by the displacement differences of the node points P_{y-1} and P_{y+1}

becomes an acceleration in the x-direction (the vice versa case also holds). In this example the points $P_{y-1,x-1}$, $P_{y-1,x+1}$, $P_{y+1,x-1}$ and $P_{y+1,x-1}$ are the cause of contraction. Their displacement in y-direction can cause a strain in the volume around the node point P_0 in a way, that this node point will get an acceleration in x-direction. The node points P_{y-1} and P_{y+1} are not important for the Poisson contraction. Their contraction can only act a force in y-direction to the node point P_0, because all three points are on a straight line in y-direction.

The acceleration than is

$$a_Q^x = \nu \frac{(y_{P_{y-1,x-1}} - y_{P_{y-1,x+1}}) - (y_{P_{y+1,x-1}} - y_{P_{y+1,x+1}})}{\Delta y^2} \tag{5.21}$$

$$a_Q^y = \nu \frac{(x_{P_{x-1,y-1}} - x_{P_{x-1,y+1}}) - (x_{P_{x+1,y-1}} - x_{P_{x+1,y+1}})}{\Delta x^2} \tag{5.22}$$

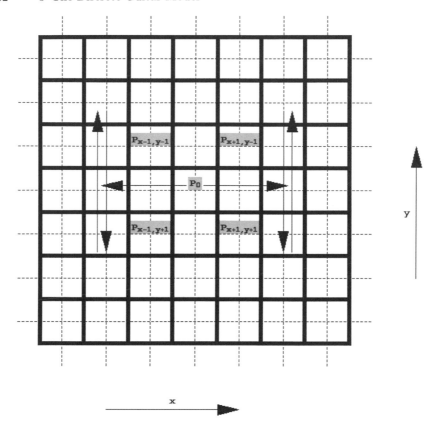

Fig. 5.4. The Poisson contraction of the point P_0 in question depends in this example of an acceleration of P_0 in x-direction on the y-displacements of the points $P_{y-1,x-1}, P_{y-1,x+1}, P_{y+1,x-1}$ and $P_{y+1,x-1}$. The points P_{y-1} and P_{y+1} play no role as their displacements in y-direction can not lead to a x-displacement of P_0

The Poisson contraction i.e. in x-direction here is modelled as the difference of the contractions of the node points $(x+1)$ and $(x-1)$ in each case with $(y-1)$ and $(y+1)$. Are the contractions on both sides equal, than the acceleration part of the Poisson contraction is zero.

5.9 Bending Waves and Their Discrete Form

The bending movements as a differential equation of fourth order use next to the point P_0 to be calculated four additional node points in each direction. This is described in Fig. 5.5.

Different from the longitudinal movement of the plate, which at each node point needs two dependent variables for the displacements in x- and in y-direction, the bending wave uses just one dependent variable z for each node

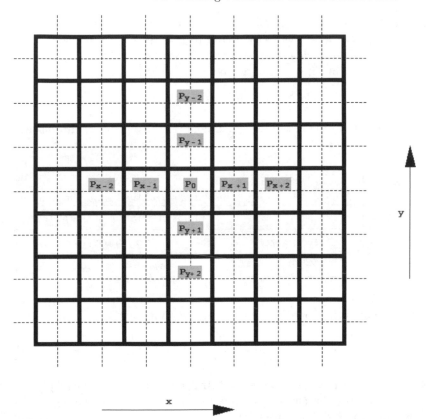

Fig. 5.5. The node points used in the calculation of the bending movements of a node point P_0 to be calculated are shown above. The bending movement only has one displacement z in direction of the plates normal vector. The acceleration of point P_0 in z-direction consists of the single accelerations from the bending of the geometry in x- and in y-direction

point for the displacement of the plate in its normal direction. The acceleration of the bending displacement now consists of the bending in the x- and y-direction like

$$a^z = a_x^z + a_y^z \,, \tag{5.23}$$

with a^z the mass-weighted overall acceleration of the bending movement for any node point and their parts a_x^z from the bending in x-direction and a_y^z from the y-direction. With the description of the single parts of the acceleration we can restrict our discussion on these components because of symmetry reasons. If we first consider the x-component, then the differential equation

$$a_x^z = \frac{\partial^2 z}{\partial t^2} = -\frac{E_x K^2}{\rho} \frac{\partial^4 z}{\partial x^4} \tag{5.24}$$

holds with z being the bending displacement at point P_0, which is differentiated two times with respect to t and four times with respect to the x-direction. Here, E_x is the Young's modulus in x-direction, ϱ is the density of the plate and K is the radius of gyration, which in the case of a plate is

$$K = \sqrt{12} \ . \tag{5.25}$$

The discrete form of the differential equation (5.24) follows from the single derivation. It holds

Variable	Order of Differentiation	Meaning
z	0	displacement of the bending movement
M	$\frac{\partial^2 z}{\partial x^2}$	radius of gyration of the plate
F_Q	$\frac{\partial M}{\partial x} = \frac{\partial^3 z}{\partial x^3}$	cross force
F_R	$\frac{\partial F_Q}{\partial x} = \frac{\partial^4 z}{\partial x^4}$	back-driving force

with the differential equations for the bending moment

$$M = E_x \frac{\partial^2 z}{\partial x^2} \int \int l^2 \, dA \tag{5.26}$$

with the distance l over and under a neutral axis through the plates plane, with is integrated over the cross-sectional area A, and with the Young's modulus E_x in x-direction, for the cross-force

$$F_Q = E_x K^2 \frac{\partial^3 z}{\partial x^3} \tag{5.27}$$

and finally for the back-driving force of the plate in x-direction

$$F_R = E_x K^2 * \frac{\partial^4 z}{\partial x^4} \ . \tag{5.28}$$

The discrete form follows the same reasoning from the radius of gyration over the cross force to the back-driving force. To obtain the back-driving force F_R, we must have two cross forces F_Q, which deviate over the grid distance like

$$F_R = \frac{F_{Q1} - F_{Q2}}{\Delta x} \ . \tag{5.29}$$

These cross forces now need again two gyration moments M for their calculation like

$$F_{Q1} = \frac{M_1 - M_2}{\Delta x} \tag{5.30}$$

$$F_{Q2} = \frac{M_2 - M_3}{\Delta x} \ . \tag{5.31}$$

The gyration moments M are the second derivation with respect to the x-direction

$$M_1 = E_x K^2 \frac{x_{P_{x-2}} - 2x_{P_{x-1}} + x_{P_0}}{\Delta x^2} \tag{5.32}$$

$$M_2 = E_x K^2 \frac{x_{P_{x-1}} - 2x_{P_0} + x_{P_{x+1}}}{\Delta x^2} \tag{5.33}$$

$$M_3 = E_x K^2 \frac{x_{P_0} - 2x_{P_{x-1}} + x_{P_{x+2}}}{\Delta x^2} \tag{5.34}$$

The mass-weights acceleration a_x^z then is

$$a_x^z = \frac{F_R}{\varrho} , \tag{5.35}$$

or after insertion

$$a_x^z = \frac{E_x K^2}{\varrho} \frac{(x_{P_{x-2}} - 2x_{P_{x-1}} + x_{P_0}) - 2(x_{P_{x-1}} - 2x_{P_0} + x_{P_{x+1}}) + (x_{P_0} - 2x_{P_{x-1}} + x_{P_{x+2}})}{\Delta x^4} , \tag{5.36}$$

or simply

$$a_x^z = \frac{E_x K^2}{\varrho} \frac{6x_{P_0} - 4x_{P_{x-2}} - 4x_{P_{x+2}} + 2x_{P_{x-1}} + 2x_{P_{x+1}}}{\Delta x^4} . \tag{5.37}$$

After calculating the acceleration from the x-bending in that way, in analogy the acceleration from the y-bending is stated by

$$a_y^z = \frac{E_y K^2}{\varrho} \frac{6y_{P_0} - 4y_{P_{y-2}} - 4y_{P_{y+2}} + 2y_{P_{y-1}} + 2y_{P_{y+1}}}{\Delta y^4} \tag{5.38}$$

and with (5.37) the overall acceleration a^z is determined.

5.10 Boundary Conditions

The boundary conditions of the Finite-Difference calculation include the cases of the Dirichlet boundary condition of a fixed boundary with u^z as displacement in x-direction like

$$u^z = 0 \quad \text{and} \quad \frac{\partial u^z}{\partial x} = 0 \quad \text{on } \Gamma \tag{5.39}$$

with Γ as boundary, and the Neumann-condition of a free boundary with

$$F_{\text{external}} = 0 \quad \text{on } \Gamma , \tag{5.40}$$

with the vanishing of external forces F_{external} on the boundary.

The theoretically possible case of a clammed boundary is not used in this book; for its treatment we point to the literature (Wagner 1947).

At the boundary two cases have to be discussed. The first case is a bending wave with a differential equation of fourth order

$$\frac{\partial^4 u^z}{\partial x^4} k_B = \frac{\partial^2 u^z}{\partial t^2} \tag{5.41}$$

with d_B as constant material values. The second case is a longitudinal wave as a differential equation of second order

$$\frac{\partial^2 u^x}{\partial x^2} k_L = \frac{\partial^2 u^x}{\partial t^2} \tag{5.42}$$

again with k_L as material constant for the longitudinal movement.

5.10.1 Boundary Conditions with Longitudinal Boundaries

The Fixed Boundary

The C^∞ steady geometry is transformed into a discrete node and element grid as described above. The least points of this grid lay on the boundary of the geometry. It holds

$$\mathbf{P}_{\text{Boundary}} = \Gamma \tag{5.43}$$

with $\mathbf{P}$ as the coordinate of the grid points

$$\mathbf{P} = \{P_x, P_y, P_z\} \tag{5.44}$$

and $\mathbf{P}_{\text{Boundary}}$ as the grid points, with lay on the boundary Γ.

The discrete form ends at the boundary. The condition of a Dirichlet boundary are fulfilled if

$$u^x = 0 \quad \text{and} \tag{5.45}$$

$$\frac{\partial u^x}{\partial x} = 0 \quad \text{and} \tag{5.46}$$

$$u^y = 0 \quad \text{and} \tag{5.47}$$

$$\frac{\partial u^y}{\partial x} = 0 \tag{5.48}$$

$$\text{for all time points } t \text{ and for all } \mathbf{P}_{\text{boundary}} \tag{5.49}$$

In the Finite-Difference model in our case of a differential equation of second order, the displacement in x- and y-direction of a node point $\mathbf{P}$ is given by

$$\frac{\partial^2 u^x}{\partial t^2} = k_L \frac{u^x_{P_{x-1}} - 2u^x_{P_0} + u^x_{P_{x+1}}}{\Delta x^2} . \tag{5.50}$$

and

$$\frac{\partial^2 u^y}{\partial t^2} = k_L \frac{u^y_{P_{y-1}} - 2u^y_{P_0} + u^y_{P_{y+1}}}{\Delta y^2} . \tag{5.51}$$

with equal distances assumed

$$\Delta x = P^x_{x+1} - P^x_0 = P^x_0 - P^x_{x-1} \qquad (5.52)$$

and

$$\Delta y = P^y_{y+1} - P^y = P^y_0 - P^y_{y-1} . \qquad (5.53)$$

So the boundary node point is used in the calculation of the acceleration of two points, first for $\mathbf{P}_{boundary-1}$, the point next to the boundary and then of course for the boundary node point itself $\mathbf{P}_{boundary}$.

The node point being two grid distances Δx away from the boundary is already excluded from a discussion of a fixed boundary in x-direction or in a grid distance Δy in analogy for the discussion of a fixed boundary in y-direction. For the acceleration calculation in each of its possible degrees of vibrational freedom this node points two grid distances away from the boundary is not touched by the boundary node point $\mathbf{P}_{boundary}$.

So the boundary node point itself is not calculated at all. It is set to zero by the boundary condition and so can not be accelerated. For this boundary node point it holds

$$\frac{\partial^2 u^x}{\partial t^2} = 0 \qquad (5.54)$$

$$\text{and } u^x = 0 \quad \text{for all } t \qquad (5.55)$$

and

$$\frac{\partial^2 u^y}{\partial t^2} = 0 \qquad (5.56)$$

$$\text{and } u^y = 0 \quad \text{for all } t \qquad (5.57)$$

It only plays a role for the point $\mathbf{P}_{boundary-1}$, means for the grid point next to the boundary in direction of calculation. If we look at a movement in x-direction and be $\mathbf{P}_{boundary}$ fixed, additionally be $\mathbf{P}$ the node point in question right next to the boundary node point. Then an equivalence of significations hold for the two standpoints, the standpoint of the boundary and the standpoint of the node point to be calculated next to the boundary as follows:

standpoint of the boundary	standpoint of the node point to be calculated
$\mathbf{P}_{boundary} =$	$\mathbf{P}_{-1}$
$\mathbf{P}_{boundary-1} =$	$\mathbf{P}$
$\mathbf{P}_{boundary-2} =$	$\mathbf{P}_{+1}$

So after insertion of the boundary condition in the differential equation for the acceleration in x-direction of the grid point $\mathbf{P}_{boundary-1}$ it is

$$\frac{\partial^2 u^x}{\partial t^2} = \frac{-2u^x_P + u^x_{P+1}}{\Delta x^2} \quad \text{for } \mathbf{P}_{boundary-1} . \qquad (5.58)$$

This is the most simple form of calculation of a boundary, as there are no additional assumptions have to be made. It is enough to keep the boundary element displacement zero, means not to calculate it at all and to perform the calculation of the neighboring boundary point without the boundary grid point.

The Free Boundary Condition

The treatment of the free boundary condition is done in two steps. In a fist step the conditions are derived for which these boundary conditions are fulfilled. In a second step the conditions are implemented and the calculation is performed. For a free boundary in x-direction it holds

$$F_{\text{external}} = 0 \ . \tag{5.59}$$

So the boundary is free of external forces. This is, because from outside of the geometry no forces can act on the boundary as there is no material left, which could cause a force. Mathematically this is a setting to zero of all terms of second derivation with respect to space, which act from outside on the boundary. It shall be remembered at the differential equation for the longitudinal movement,

$$\frac{\partial^2 y}{dx^2} \, k_L = \frac{\partial^2 y}{\partial t^2} = a^x \ , \tag{5.60}$$

with the acceleration a^x equivalent to the second derivation with respect to space $\frac{\partial^2 y}{\partial t^2}$ and the material constant k_L. This acceleration a then is the cause of the back-driving force of the grid point $F = m \, a$ as well as the cause of its change of movement (like discussed in the previous section about the discrete form of the single parts of the guitar).

If we use this terminology of node points like we introduced them in the section of the fixed boundary of the longitudinal movement above, for the acceleration of the boundary element $a_{\mathbf{P}_{\text{Boundary}}}$ holds

$$a_{\mathbf{P}_{\text{Boundary}}} = \left(\frac{d^2 y}{dx^2} \right)_{\text{internal}} + F_{\text{external}} \quad \text{with} \quad F_{\text{external}} = 0 \tag{5.61}$$

We must keep in mind now, that the acceleration of the boundary point like above has to come out of three points, but now not out of $\mathbf{P}_{\text{Boundary}}$, $\mathbf{P}_{\text{Boundary}-1}$ and $\mathbf{P}_{\text{Boundary}-2}$, but out of $\mathbf{P}_{\text{Boundary}}$, $\mathbf{P}_{\text{Boundary}+1}$ and $\mathbf{P}_{\text{Boundary}-1}$. Different from the fixed boundary, $\mathbf{P}_{\text{Boundary}}$ is included in the calculation now. Therefore the virtual node point $\mathbf{P}_{\text{Boundary}+1}$ is introduced with the coordinates

$$\mathbf{P}^x_{\text{Boundary}+1} = \mathbf{P}^x_{\text{Boundary}} + (\mathbf{P}^x_{\text{Boundary}} - \mathbf{P}^x_{\text{Boundary}-1}) = \mathbf{P}^x_{\text{Boundary}} + \Delta x \ . \tag{5.62}$$

The distance of the virtual node point outside of the geometry to the boundary node point is again the grid point distance Δx. The grid is enlarged at the boundary with a virtual element.

The boundary grid point indeed is accelerated by the guitar geometry, but not from the additional virtual boundary node point. The additional amount of acceleration of the virtual grid point $P_{Boundary+1}$ on the boundary point $P_{Boundary}$ must vanish, but the additional value of the next grid point in the geometry, here in x-direction $P_{Boundary-1}$, will further on influence the movement of the boundary grid point. If we write the difference equation down

$$a = k_L \frac{\frac{u^x_{Boundary+1} - u^x_{Boundary}}{\Delta x} - \frac{u^x_{Boundary} - u^x_{Boundary+1}}{\Delta x}}{\Delta x} \tag{5.63}$$

and demand that the influence of the boundary on the acceleration of the boundary grid points must be zero

$$\frac{u^x_{Boundary+1} - u^x_{Boundary}}{\Delta x} = 0 \tag{5.64}$$

it follows, that

$$u^x_{Boundary+1} - u^x_{Boundary} = 0 \tag{5.65}$$

or

$$u^x_{Boundary+1} = u^x_{Boundary} \tag{5.66}$$

or for the acceleration of the boundary grid points in x-direction

$$a = k_L \frac{u^x_{Boundary} - u^x_{Boundary+1}}{\Delta x^2} \tag{5.67}$$

for every point in time t.

This is not very useful for an implementation as two kinds of conditions have to be checked. First it has to be realized, if the point in treatment is a boundary point, then if the boundary condition is a fixed or a free one. In praxis the idea has proven to work better to really implement the virtual node point and after each temporal iteration step of the calculation for all virtual grid points to perform the equality condition mentioned above.

5.10.2 Boundary Condition of Bending Movements

The bending movement is a differential equation of fourth order. It holds, for the sake of plasticity just shown in x-direction here, for a rod

$$\frac{\partial^4 u^z}{\partial x^4} k_B = \frac{\partial^4 u^z}{\partial t^4} \tag{5.68}$$

Again, we will discuss the two cases of a fixed and a free end. Like above, because of reduced calculation cost, an implementation of virtual grid points is preferred here. Different from the longitudinal movement the discrete implementation of the bending waves need five node points. For the boundary node point they are

$\mathbf{P}_{\text{Boundary}-2}$: Grid point in geometry two points away from the boundary grid point

$\mathbf{P}_{\text{Boundary}-1}$: Grid point in geometry next to boundary grid point

$\mathbf{P}_{\text{Boundary}}$:   Boundary grid point

$\mathbf{P}_{\text{Boundary}+1}$: virtual grid point next to boundary grid point

$\mathbf{P}_{\text{Boundary}+2}$: virtual grid point two point away from boundary grid point

Out of these five points three bending moments are calculated (see also discrete form of the differential equation of the bending movement in the previous section)

$$M_- = k_M \frac{u^z_{\text{Boundary}-2} - 2u^z_{\text{Boundary}-1} + u^z_{\text{Boundary}}}{\Delta x^2} \qquad (5.69)$$

with $\mathbf{P}_{\text{Boundary}}$, $\mathbf{P}_{\text{Boundary}-1}$, $\mathbf{P}_{\text{Boundary}-2}$

$$M = k_M \frac{u^z_{\text{Boundary}-1} - 2u^z_{\text{Boundary}} + u^z_{\text{Boundary}+1}}{\Delta x^2} \qquad (5.70)$$

with $\mathbf{P}_{\text{Boundary}+1}$, $\mathbf{P}_{\text{Boundary}}$, $\mathbf{P}_{\text{Boundary}-1}$

$$M_+ = k_M \frac{u^z_{\text{Boundary}} - 2u^z_{\text{Boundary}+1} + u^z_{\text{Boundary}+2}}{\Delta x^2} \qquad (5.71)$$

with $\mathbf{P}_{\text{Boundary}+2}$, $\mathbf{P}_{\text{Boundary}+1}$, $\mathbf{P}_{\text{Boundary}}$.

From these bending moments the accelerations of the boundary grid points are calculated like discussed in this text in the chapter about the discrete forms of the bending movements. There, the bending moment M is mentioned, which is the second derivation with respect to space with K_M being the material constant of the bending moment. Out of this, the acceleration of the boundary element follows as

$$\frac{\partial^2 u^{z \text{ just in x-direction}}}{\partial t^2} = -k_B \frac{M_- - 2M + M_+}{\Delta x^2} \qquad (5.72)$$

A further discussion of the discrete form of the bending equation is of no use here, because from the bending moments the further steps can be seen quite clear. We can remember here, what in the above section about the boundary conditions of the longitudinal movements has been shown. Because the amount of acceleration coming from the virtual grid points through the moments enter the equation of the bending movement acceleration in a linear way, we can directly look at the points $\mathbf{P}$ and their displacement in z-direction.

Fixed Boundary Condition with Bending Movements

If we look at the fixed boundary, we fist have to mention that the equation

$$u^z = 0 \quad \text{on } \Gamma \tag{5.73}$$

is exactly fulfilled, if

$$u^z_{\text{Boundary}} = 0 \quad \text{for all } t \tag{5.74}$$

The two virtual grid points beyond the boundary not have to be taken into consideration. Still they are implemented because of the above mentioned speeding up of calculation time. So it also holds that

$$u^z_{\text{Boundary}+1} = 0 \quad \text{for all } t \tag{5.75}$$

and

$$u^z_{\text{Boundary}+2} = 0 \quad \text{for all } t \ . \tag{5.76}$$

Therefore

$$M_+ = 0 \tag{5.77}$$

and

$$M = \frac{u^z_{\text{Boundary}-1} - 2u^z_{\text{Boundary}}}{\Delta x^2} \ . \tag{5.78}$$

So it not only is ensured, that

$$u^z_{\text{Boundary}} = 0 \quad \text{for all } t \tag{5.79}$$

but also, that

$$\frac{\partial u^z}{\partial x_{\text{Boundary}}} = 0 \quad \text{for all } t \ , \tag{5.80}$$

which also means, that the slope of the boundary is always zero, so the boundary is not clammed but indeed fixed. Here it follows out of the simple consideration, that

$$\frac{\partial u^z}{\partial x_{\text{Boundary}}} = \frac{u^z_{\text{Boundary}+1} - u^z_{\text{Boundary}}}{\Delta x} = 0 \ , \tag{5.81}$$

due to

$$u^z_{\text{Boundary}} = 0 \quad \text{and} \quad u^z_{\text{Boundary}+1} = 0 \ , \tag{5.82}$$

that the slope of the boundary for all t is zero, too.

The Free Boundary in the Differential Equation of the Bending Movement

The free boundary of the differential equation of the bending movement follows in analogy to the considerations of the free boundary with the longitudinal movement. Outside the boundary in the region of the virtual grid points, no acceleration shall take place.

$$F_{\text{external}} = 0 \quad \text{on } \Gamma \tag{5.83}$$

This is realized again with a similar method of virtual grid points. The missing of an acceleration means first, that

$$M_+ = \frac{u^z_{\text{Boundary}} - 2u^z_{\text{Boundary}+1} + u^z_{\text{Boundary}+2}}{\Delta x^2} = 0 , \tag{5.84}$$

then also it requires, that

$$M = \frac{\frac{u^z_{\text{Boundary}-1} - u^z_{\text{Boundary}}}{\Delta x} + \frac{u^z_{\text{Boundary}} - u^z_{\text{Boundary}+1}}{\Delta x}}{\Delta x} \tag{5.85}$$

becomes

$$M = \frac{\frac{u^z_{\text{Boundary}-1} - u^z_{\text{Boundary}}}{\Delta x}}{\Delta x} . \tag{5.86}$$

The reason here is, that the acceleration does not depend on the virtual grid points. Therefore the condition after equating (5.79) and (5.80) is

$$0 = \frac{u^z_{\text{Boundary}} - u^z_{\text{Boundary}+1}}{\Delta x^2} \tag{5.87}$$

or simply

$$u^z_{\text{Boundary}} = u^z_{\text{Boundary}+1} \tag{5.88}$$

and after insertion of (5.88) in (5.85)

$$u^z_{\text{Boundary}+1} = u^z_{\text{Boundary}+2} . \tag{5.89}$$

The condition of a free boundary is also obviously fulfilled, if

$$u^z_{\text{Boundary}} = u^z_{\text{Boundary}+1} = u^z_{\text{Boundary}+2} . \tag{5.90}$$

This system is implemented on a computer by equalling the z-values of the virtual boundary points and the z-values of the boundary points for each time step. So for each time step of the calculation of the bending movement it is ensured, that the acceleration at the boundary is zero, so we have a vonNeumann boundary condition of a free boundary.

6

Bending, Damping and Coupling

The guitar is a system made up of several coupled components: top plate, back plate, ribs, neck and interior air space. There are several ways of tackling the contact problems at the glued joints between these components. The following discussion of the couplings includes consideration of the boundary conditions, which are known in the literature as a mixture of fixed and clammed boundaries. The guitar top plate will serve as an example. The top plate is glued to the rims all around its boundary; this comes close to a fixed boundary condition. But it also is coupled at this boundary to the ribs, so in a certain range free to vibrate. These special conditions known from the guitar geometry are modelled in a way, that the boundary is fixed, but the last element of the top plate which lies on the ribs is free to vibrate and to couple with the ribs. The only exception of this boundary condition is the neck of the guitar, which is free at its sides and just fixed at the glue point to the top plate.

Besides the couplings between the guitar elements, couplings between longitudinal and bending movements occur. These couplings are met at the places of the guitar geometry, where there are curved areas and are treated separately. The ribs is curved by a hot bending procedure and therefor nearly no stresses occur in its structure because of this bending. The curved structure of the ribs produce a steady transformation of bending waves into longitudinal movements and vice versa. The guitar back plate on the other side has its slight bending not through the above mentioned hot bending technique of the ribs. It is crafted a little bit bigger than the ribs rim, so when attached into that ribs rim, it curves a little bit in its normal vector direction. The curvature of the top plate is just a consequence of the coupling between longitudinal and bending stresses and strains within the back plate. The two curvatures mentioned above (geometrical curvature with the ribs, stress-strain curvature with the back plate) are of different nature and so have to be mathematically treated different.

Another factor determining the guitar sound is the damping of the different guitar parts. Mathematically, damping is normally formulated as an additional term in dependence of the velocity. But this leads to a change in the eigenvalues of the vibrating system in dependence of the strength of the damping, what can cause problems in the modelling of the guitar parts; here i.e. are the changes of the resonance behavior of different sorts of wood, temperature, degree of moist have to be taken into consideration. So here a kind of damping is proposed, with does not effect the eigenvalues of the plates, but for which superposition still holds. This kind of damping is physically-geometrically plausible and therefore indeed has to be demanded. Nevertheless, the usual kind of damping is still used as a velocity term in the differential equation of movement, but which is just used with small values (for the reasons of using both methods see text below).

6.1 Coupling of Longitudinal and Bending Waves

As already indicated in the section on the discrete formulation of bending waves, curved structures can be described in terms of the coupling between longitudinal and transverse vibrations – in this case bending vibrations. Two methods are used here. The first is a coupling in which the longitudinal compressions of the plate affect the transverse compressions if the plate is already under tension. The second type of coupling described here can be used if the bending is already present in the structure, as is the case in the ribs of the guitar. But even then there are longitudinal and transverse couplings, which allow the movement to occur.

For the sake of simplicity, we shall use the terms "first type coupling" for the first coupling described above, and "second type coupling" for the second.

6.1.1 LT – First Type Coupling

For the coupling it is necessary, that the accelerations of longitudinal and bending waves get an additional coupling term. So (5.16) and (5.17) become

$$a^x = a^x_{L,K} + a^x_{S,K} + a^x_{Q,K} \tag{6.1}$$

$$a^y = a^y_{L,K} + a^y_{S,K} + a^y_{K,Q} \tag{6.2}$$

and (5.23) becomes

$$a^z = a^z_{x,K} + a^z_{y,K} \tag{6.3}$$

where all acceleration terms are now coupling-weighted with the x- and y-parts of the longitudinal movement (or in-plane movement) $\mathbf{a}_{L,K} = \{a^x_{L,K}, a^y_{L,K}\}$, as shearing $\mathbf{a}_{S,K} = \{a^x_{S,K}, a^y_{S,K}\}$, as Poisson contraction $\mathbf{a}_{Q,K} = \{a^x_{Q,K}, a^y_{Q,K}\}$ and the x- and y-parts of the bending movement $\mathbf{a}^z_K = \{a^z_{x,K}, a^z_{y,K}\}$.

The coupling is caused by the fact, that the differences in each kind of movement takes place in the direction, in with the other kind of movement also occurs and so the axis of the differentials are compressed or extended. Looking at the fundamental equation of the longitudinal movement (5.1) (now without shearing and Poisson contraction for simplicity, but for which the conditions are analog), the change of length in the x- and y-direction respectively is

$$\Delta x^{P_0}(t) = \sqrt{\Delta x_0^2 + \left(\frac{z_{P_{x+1}} - z_{P_{x-1}}}{2}\right)^2} \qquad (6.4)$$

$$\Delta y^{P_0}(t) = \sqrt{\Delta y_0^2 + \left(\frac{z_{P_{y+1}} - z_{P_{y-1}}}{2}\right)^2} \qquad (6.5)$$

with Δx_0 as grid constant in x-direction at point P_0, and $\Delta x^{P_0}(t)$ grid constant at point P_0 at the time point of calculation. This is prolonged, if the displacement of the grid points in x-direction around $z_{P_{x+1}}$ and $z_{P_{x-1}}$ and around $z_{P_{y+1}}$ and $z_{P_{y-1}}$ are different. In this case the grid constant or grid distance at point P_0 changes. This acts on the acceleration in the x- and y-direction of the longitudinal waves.

In Fig. 6.1 the case of a coupling of a longitudinal movement, here in x-direction on a bending movement, also in x-direction is shown. The

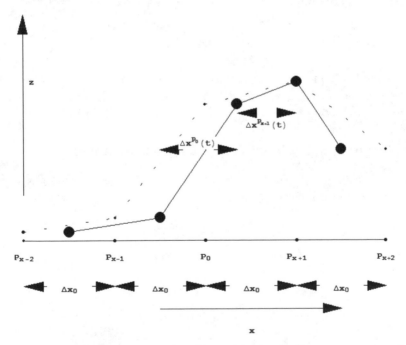

Fig. 6.1. Schematic figure of the coupling of longitudinal movement on a bending movement. The distance of the grid points (grid constant) Δx_0, for which the bending differential equation in the discrete case is differentiated, is in the case of a coupling of a longitudinal wave at each grid point changed from the original values. So the longitudinal movement (here in x-direction) has influence on the acceleration of the bending movement corresponding of its time dependent grid distance $\Delta x^{P_0}(t)$ or $\Delta x^{P_{x+1}}(t)$, which are shown here as an example

z-displacement of the grid points in both curves, the dotted and the solid, are equal. In the case of the solid bending line, the longitudinal displacements in x-direction are taken into consideration, too. The grid distances have changed here considerably. In the case of a coupling of the longitudinal movement in the bending wave, because here five points are included in the calculation, a calculation as with the longitudinal case is very difficult. Here, the coupling is directly performed in the calculation of the bending moment, the cross-force and the backdriving force.

The differential equation of the bending movement with coupling of longitudinal movements is

$$\frac{\partial^2 u_z}{\partial t^2} = -\frac{E * K^2}{\rho} \left(\frac{\partial^4 u_z}{\partial \left(x + \frac{\partial^2 u_x}{\partial x^2} \right)^4} + \frac{\partial^4 u_z}{\partial \left(y + \frac{\partial^2 u_y}{\partial y^2} \right)^4} \right) . \tag{6.6}$$

Here, the bending in x- and y-direction is summarized. In the discrete form all terms which have a grid distance Δx or Δy respectively change according to the longitudinal impact. The differential

$$\partial \left(x + \frac{\partial^2 u_x}{\partial x^2} \right)^4 \tag{6.7}$$

is to be understood in a way, that the normally homogeneous x-axis is now strained by the longitudinal differential $\frac{\partial^2 u_x}{\partial x^2}$. For the differential equation this means the inclusion of the new grid distances Δx^P and Δy^P.

Coupling of Bending Waves into Longitudinal Waves

Starting with the differential equation of second order for the longitudinal movements (5.1) becomes

$$a = k_S \frac{\frac{s(n-1)-s(n)}{\Delta x^{P-1}} - \frac{s(n)-s(n+1)}{\Delta x^{P+1}}}{(\Delta x^{P-1} + \Delta x^{P+1})/2} \tag{6.8}$$

with

$$\Delta x^{P-1} = \sqrt{\Delta x_0^2 + \left(\frac{z_{P_{x-1}} - z_{P_x}}{2} \right)^2} \tag{6.9}$$

$$\Delta x^{P+1} = \sqrt{\Delta x_0^2 + \left(\frac{z_{P_x} - z_{P_{x+1}}}{2} \right)^2} \tag{6.10}$$

and Δx_0 as the unchanged grid distance. The coupling in the plate movement in relation to the shearing and Poisson contraction follows in the same way.

Coupling of the Longitudinal Waves into the Bending Waves

The case of the coupling of longitudinal movements into the bending movements is analog to the vice versa case of a coupling of bending waves into longitudinal waves. But here, vier altered grid distances occur like

$$\Delta x^{P-2} = \sqrt{\Delta x_0^2 + \left(\frac{z_{P_{x-2}} - z_{P_{x-1}}}{2}\right)^2} \tag{6.11}$$

$$\Delta x^{P-1} = \sqrt{\Delta x_0^2 + \left(\frac{z_{P_{x-1}} - z_{P_x}}{2}\right)^2} \tag{6.12}$$

$$\Delta x^{P+1} = \sqrt{\Delta x_0^2 + \left(\frac{z_{P_x} - z_{P_{x+1}}}{2}\right)^2} \tag{6.13}$$

$$\Delta x^{P+2} = \sqrt{\Delta x_0^2 + \left(\frac{z_{P_{x+1}} - z_{P_{x+2}}}{2}\right)^2}. \tag{6.14}$$

So (5.36) becomes

$$a_{x,\text{coupled}}^z = \frac{E_x K^2}{\varrho} f \tag{6.15}$$

with

$$f =$$

$$\frac{\dfrac{\dfrac{\dfrac{s(n-2)-s(n-1)}{\Delta x^{P-2}} - \dfrac{s(n-1)-s(n)}{\Delta x^{P-1}}}{(\Delta x^{P-2}+\Delta x^{P-1})/2} - \dfrac{\dfrac{s(n-1)-s(n)}{\Delta x^{P-1}} - \dfrac{s(n)-s(n+1)}{\Delta x^{P+1}}}{(\Delta x^{P-1}+\Delta x^{P+1})/2}}{(\Delta x^{P-2}+\Delta x^{P-1}+\Delta x^{P+1})/3} - \dfrac{\dfrac{\dfrac{s(n-1)-s(n)}{\Delta x^{P-1}} - \dfrac{s(n)-s(n+1)}{\Delta x^{P+1}}}{(\Delta x^{P-1}+\Delta x^{P+1})/2} - \dfrac{\dfrac{s(n)-s(n+1)}{\Delta x^{P+1}} - \dfrac{s(n+1)-s(n+2)}{\Delta x^{P+2}}}{(\Delta x^{P+1}+\Delta x^{P+2})/2}}{(\Delta x^{P-1}+\Delta x^{P+1}+\Delta x^{P+2})/3}}{(\Delta x^{P-2} + \Delta x^{P-1} + \Delta x^{P+1} + \Delta x^{P+2})/4}. \tag{6.16}$$

Equation (6.16) follows the same principle like (6.8) for the longitudinal movement, with the difference, that in (6.16) we have to consider the fourth order. The equation above for the bending movements in x-direction hold equivocally for the y-direction and it holds like in the uncoupled case

$$a_{\text{coupled}}^z = a_{x,\text{coupled}}^z + a_{y,\text{coupled}}^z. \tag{6.17}$$

So now the coupling of longitudinal movements into bending movements and vice versa the coupling of bending waves into longitudinal waves is described. This coupling takes place in all structures of the guitar, which have bending and longitudinal waves, in the top and back plate, the ribs and the neck. All there the one kind of movement will occur as soon as the other kind of movement is induced. I.e. the top plate will vibrate in longitudinal waves, although it is moved by the strings only in bending direction. This model taken from the plates real movements are the cause for additional phenomena, like i.e. surface waves, which can be watched in the simulation as will (see results chapter).

This type of coupling describes the curvature of the back plate. This plate is made somewhat larger than the ribs. Although there are several techniques used by guitar builders to construct the curvature of the back plate, mathematically we have it squashed at the sides. This squashing corresponds to a slight bulge in the back plate. It is not carved to shape, like the top and back of a violin, for example; instead it is in a state of constant tension. This tension exists in the plane of the plate. Theoretically, the plate does not have to bulge, but could instead remain in a flat, compressed state. But there are two things that prevent this. First, the back plate is made approximately two millimeters too big. The force that would be necessary to squash such a back plate in between the ribs would actually break them. The other factor is the coupling between the longitudinal and transverse movements of the plate.

The force applied to the back plate is a longitudinal one. The resulting bulge is transverse. Obviously, the force is being passed on in some form in the plate. This conversion can easily be imagined if we consider one element of the plate. This element is compressed longitudinally. If we now displace it a little in the transverse direction, for example by knocking on the back plate, then a curvature arises at the location of this element. This curvature becomes a wave that runs along the plate. This is the case here. But here we also find that the longitudinal compression has the possibility of spreading upwards. For the greater the curvature of the top plate, either locally or overall, the longer it becomes in the longitudinal direction. This results in the plate relaxing, since it has more room for accommodating the compression. This spread, however, is hampered by the bulge that arises. For the bending created at the observed element leads to a restoring force, which counteracts the displacement caused by the longitudinal squashing. And thus a state of equilibrium arises, in which the curvature is so large that the restoring force of the curvature equals the bulging force of the longitudinal squashing.

$$\int_B k_B \left(\frac{\partial^4 z_{P_B}}{\partial x^4} + \frac{\partial^4 z_{P_B}}{\partial y^4} \right) = -\int_B k_L \left(\frac{\partial^2 x_{P_B}}{\partial x^2} + \frac{\partial^2 y_{P_B}}{\partial y^2} \right) \qquad (6.18)$$

Figure 6.2 shows the distended back plate. The position of the bracing has not been shown at equilibrium, to give greater clarity in the three-dimensional representation. The uniform shape of the distension was reached from an elastic modulus ratio of around $E_x/E_y = 8.0$. Previously a considerably smaller distension had been visible in the central axis in the area of the central bracing, resulting in a type of saddle.

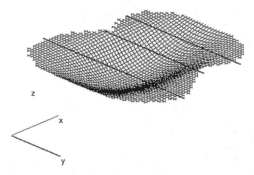

Fig. 6.2. The bulging of the back plate owing to a longitudinal force acting at the boundaries of the ribs. The hollow is created because the back plate is 2 mm larger than the space for it provided by the ribs. The transient calculation was started. Then pressure was applied virtually to the back, i.e. a particular element was displaced. With strong damping, the back plate rapidly returned to its state of equilibrium as shown here

Figure 6.3 gives an insight into the spectral conditions in the two cases. In each case a knocking sound is made, which dies away rapidly, in order to create a plausible transient. The spectra of these knocking sounds exhibit a shift of the mode frequencies of the back plate when the distension is present, which is shown in Table 6.1 for the first five modes.

Table 6.1 shows that the shift increases as the frequencies become higher. This was to be expected. If one bends a piece of wood, it clearly demonstrates audibly higher eigenfrequencies. This rise in pitch is reinforced in the spectrum as one looks at higher frequencies.

The reason for this is quite obvious. Since the longitudinal pressure causes transverse bending, every transverse movement must not only overcome the resistance to curvature, but must also overcome that part of the force that comes from the lateral squashing. This latter part must be added to the restoring force.

$$\int_B k_B \left(\frac{\partial^4 z_{P_B}}{\partial x^4} + \frac{\partial^4 z_{P_B}}{\partial y^4} \right) + \int_B k_L \left(\frac{\partial^2 x_{P_B}}{\partial x^2} + \frac{\partial^2 y_{P_B}}{\partial y^2} \right) = \int_B \frac{\partial^2 z_{P_B}}{\partial t^2} \quad (6.19)$$

This also results in the mode frequencies of the back plate becoming higher.

This would explain the enhanced brightness in tone that guitar-makers report. On the other hand, it could also lead to increased amplitudes at the higher frequencies. At first the spectra of the two knocks on the back appear to speak a different language. The spectrum without the distension seems to be slightly richer than the one with the distension.

However, perception experiments for the two sounds created leave no doubt. Not only is the increase in the modes clearly heard, but also the brightness is obviously enhanced when the back plate is distended. The reason for this can be found in the images of the spectral centroids. The peak value of

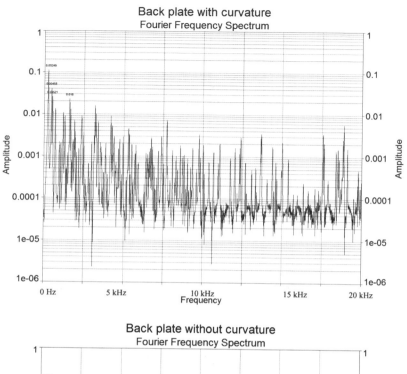

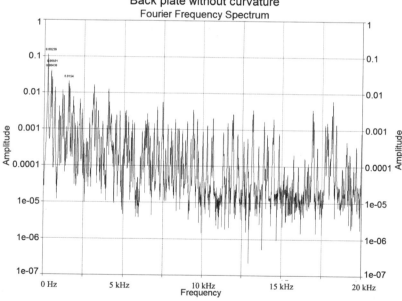

Fig. 6.3. Spectra of two knocking sounds from the back plate up to 20 kHz; (**a**) back plate distended (**b**) back plate not distended. The frequencies are shifted upwards when the distension is present (see Table 6.1). Spectrum (a) appears to be richer in the overtone spectrum than (b) (please note the amplitude scales). The peak values of the spectral centroid are higher in (a) (see next Fig. 6.4)

Table 6.1. Eigenfrequency values of the back plate in the distended and the non-distended state. The distension, which is caused by the back plate being about 2 mm larger than the ribs, leads to a shift in the eigenvalues of the back plate, which increase towards the higher frequencies

Mode	Back Plate not Distended	Back Plate Distended
1	229.44	239.04
2	420.48	439.68
3	480.96	500.16
4	652.8	681.6
5	912.0	931.2

the tone in the distended state reaches 1430 Hz and is therefore significantly higher than that in the non-distended state, at 905 Hz. The lower diagram of the spectral centroid in the non-distended state appears richer, but this must be seen relative to the highest value. Considering this, one finds that even after a short period of about 10 ms, the spectral values no longer differ significantly, although at the beginning the distended back plate has a larger proportion of high frequencies.

Here one can consider something like a coupling of the longitudinal vibrations with the transverse ones in the compressed case, in which the longitudinal vibrations couple to the transverse ones much better than they do in the case of a non-compressed back plate. Since the longitudinal eigenvalues are much higher than those of bending waves, they would brighten the sound. This can only occur if the bending waves vibrate with the mode frequencies of the longitudinal waves, since only the transverse movements of the back plate can be radiated away to any significant degree. This means that the transverse movement of the back plate is not only determined by its own modal frequencies, but also by those of its longitudinal waves. The principle behind this has already been discussed in the chapter on nonlinear couplings.

In addition, statements by guitar-makers that a back plate under tension is brighter than one which is not, can only be confirmed by investigations of this type. This is because a guitar-maker either knocks on the wood or plays a tone. The tense back plate really does sound brighter when knocked. It is possible to play two guitars, one with, and the other without, a compressed back plate. This procedure does not produce absolute proofs, because the sounds that are produced are of course influenced by many other factors. But here, too, one would expect the back plate to make the sound brighter during the initial transient, and guitar-makers and professional musicians confirm that this is so.

So, it we have demonstrated that the type of coupling between longitudinal and transverse waves does not only simulate the distension of the back plate correctly, but also leads to the properties of a back plate that is distended due to applied external pressure.

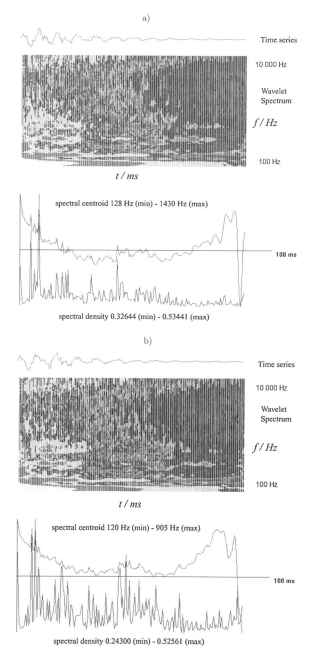

Fig. 6.4. Wavelet spectra, spectral centroids and spectral densities of two knocking sounds from the back plate calculated from 100 Hz to 10 kHz; (**a**) back plate distended (**b**) back plate not distended. The maximum and minimum values of the knocking sounds from the distended back plate are considerably higher than those of the plate when it is not distended

The second type of curved structure is that of the pre-formed bending, which is not caused and maintained by pressure. This is the state in the ribs. This will now be investigated as second type coupling.

6.1.2 LT – Second Type Coupling

The second method of coupling longitudinal and transverse movements is necessary in the ribs. An exact method is also required if other instruments, such as the violin, cello or double bass are to be iterated. This is because the curved structures of these instruments are very complicated and for this reason an exact solution is to be recommended.

The basic idea can be also expressed as a question: does a curved plate vibrate at the same frequencies as a flat plate of the same size, or do the eigenvalues change? Both the curved and the non-curved plate vibrate around a state of equilibrium. So the question is whether the curvature really influences the mode frequencies of the plate, or whether a state of equilibrium without tension is sufficient for us to speak of constant eigenfrequencies independent of the plates geometry.

The question can be answered, at least rudimentarily, using a simple experiment. If we knock on a very thin plank of wood using a finger or a fingernail, with a little practice we can discern its deepest and thus loudest eigenvalue from the sound, just by listening. This method is also used by instrument-makers. Guitar-makers (and also violin-makers, for example) use these "knocking-sounds" to match the top plate and the back plate. This works because the spectra of the plates are inharmonic. In the case of an harmonic spectrum we would get a tonal fusion. This perception would perceive the tone as a gestalt. Singe tones could not be separated (on tone fusion see i.e. (Stumpf 1883) reviewed in (Schneider 1997b); for the aesthetic notion of union in a manifold with sounds see (Lipps 1903).

If we now wedge this thin plate against a firm object such as a table, and press on the free side of the plate so that the plate bends, if we knock on it again we hear that the tone produced rises in pitch. If we heat-treat the plate and get it to take up this form without the application of tension, the eigenvalue of the plate again becomes different from that in the non-curved state. So obviously the eigenvalues of a plate depend on its geometry (and on its tension). The question is now, how this can be modelled for a curvature? The case under tension has already been discussed in the section about the first type coupling.

The reasoning goes like this: at every point the curved plate has a certain radius of curvature r. Here we have two extreme cases i.e. $r = \infty$, where the plate is flat, and $r = \Delta x$, where the plate has a bend of 90 degrees and the discretization width on x is Δx hat. In the first case the discussion is superfluous. So we will tackle below the finite radii $\infty > r > \delta x$.

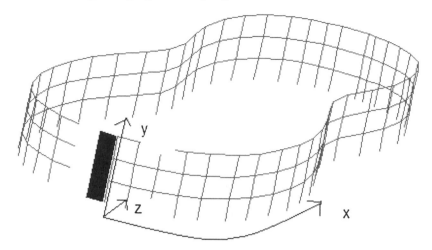

Fig. 6.5. Coordinate system of the ribs rolled out flat like a plate. The x axis is curved, so that every rib element has its own curvature and its own coordinate system with the base vectors $\mathbf{Z}^{\mathrm{Achsen}}(i)$ and the transformation tensor $\mathbf{T}^{i,i+1}$ hat

Now, as mentioned above, the definition of the radius for small values is

$$\frac{1}{r} = \frac{\partial^2 y}{\partial x^2} \tag{6.20}$$

In this case the x is discrete and depends on the geometry of the ribs at the relevant element.

The curvature is defined so that if the plate is flat the curvature is zero, and if there is a 90-degree bend, the curvature is $\pi/4$. So a simple conversion suffices.

$$k(i) = \arctan \frac{Z^y(i+1) - Z^y(i)}{Z^x(i+1) - Z^x(i)} \tag{6.21}$$

$$- \arctan \frac{Z^y(i) - Z^y(i-1)}{Z^x(i) - Z^x(i-1)} \tag{6.22}$$

Here we are speaking of the i-th rib node point which has the global coordinates (see Fig. 6.7)

$$0\,\mathrm{cm} < Z^x(i) < 49.3\,\mathrm{cm} \tag{6.23}$$

$$-13.5\,\mathrm{cm} < Z^y(i) < 13.5\,\mathrm{cm} \tag{6.24}$$

The ribs form a plate with n times m elements. Its x coordinate runs along the top plate (or the back plate), and so is actually curved. But the ribs only "know" about this via their geometry, which determines the curvatures and the base vectors. The y coordinate of the ribs is parallel to the z coordinate of the guitar (for a comparison of the axis between top plate (back plate, neck

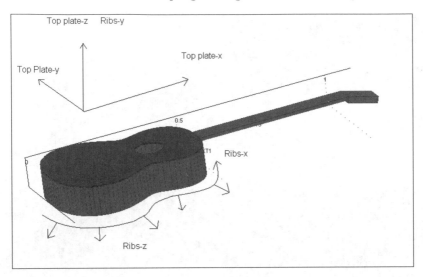

Fig. 6.6. Comparison of the coordinate systems of the top plate (back plate, neck and inclosed air of the guitar respectively) with the coordinate system of the ribs. Along the z-coordinate of the top plate (called top plate-z in the picture) is the same as the coordinate axis of the ribs (top plate-y), as the x-axis of the ribs runs around the top plate and so is different with respect to the axis direction of the top plate at each rib point. So the top plate-x and top plate-z in the picture can not be assigned to the top plate axis

and inclosed air respectively) and the ribs see Fig. 6.6). Along the y-axis of the ribs, all curvatures are equal. Along Z^x they will change for each node point i along the ribs.

Each node point of the ribs can now be assigned to a coordinate system, which always has the z-axis of the top plate as its y-axs, a normal vector outwards at the ribs point to the z-axis and a tangent vector at the ribs point in the x/y plane of the top plate to the x-axis. The base vectors of each individual element are equal to the base vectors of every other guitar element, since every element naturally exists in its own "straight world". This holds for all base vectors $\mathbf{Z}^{\mathrm{Achsen}}(i)$ of the node points i

$$\mathbf{Z}^{\mathrm{Axis}}(i) = \begin{bmatrix} 1 & 0 & 0 \\ 0 & 1 & 0 \\ 0 & 0 & 1 \end{bmatrix}. \tag{6.25}$$

As each rib point coordinate system lied differently in space, we need transformation matrices between two neighbouring ribs coordinate systems (see Fig. 6.5) defined as

$$\mathbf{T}^{i,i+1} = \begin{bmatrix} \cos(k^i + k^{i+1}) & 0 & 0 \\ 0 & 1 & 0 \\ 0 & 0 & 1 \end{bmatrix}, \tag{6.26}$$

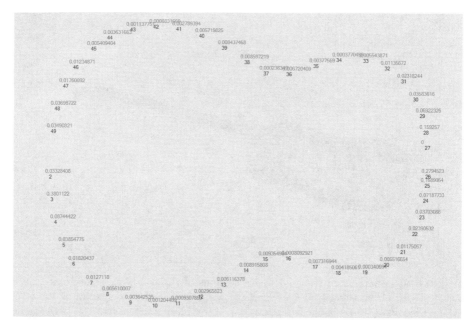

Fig. 6.7. Coordinate system of the ribs rolled out flat like a plate. The x axis is curved, so that every rib element has its own curvature and its own coordinate system with the base vectors $\mathbf{T}^{i,i+1}$, and the transformation tensor $k(i)$

so that

$$\mathbf{Z}(i + 1) = \mathbf{T}^{i,i+1}\, \mathbf{Z}(i) \,. \tag{6.27}$$

The expression $\cos(k^i + k^{i+1})$ does not only mean that this transformation matrix reduces to one, if both curvatures k^i and k^{i+1} are zero. The expression is also equal to zero if both elements are curved to the same degree away from each other. In the guitar this is indeed the case when the large curvature of the lower body turns into a curvature in the opposite direction at the waist. There the curvature is equal to zero at a turning point.

If we now put the above radius of curvature into the differential equation for the bending wave, our fourth derivative must be multiplied by the second, i.e. by the curvature. Just on the second derivation, which at each point of the curved ribs geometry is necessary a different value, the bending movement takes its cause. The first two derivations are due to the geometry, which we first have to eliminate to calculate the pure bending waves. So get a derivation of sixth order with a factor P_C for the curvature at each node point.

$$\frac{\partial^2 u_z}{\partial t^2} = -\frac{E * K^2}{\rho}\left(P_C \frac{\partial^6 u_z}{\partial x^6} + \frac{\partial^4 u_z}{\partial y^4}\right) \tag{6.28}$$

It should be noted that this equation is only valid for one element, since every element has its own curvature. And the curvature is only found in the x derivative, since the y coordinate of the ribs lies in the z direction of the guitar, where there is no curvature.

Unfortunately we cannot neglect the curvature and look at the ribs as a flat plate. So de facto with a curved rib element, the restoring force is weakened in the local z direction. This is because the curvature weakens the x element in the transformation matrix, i.e. the coupling to the neighboring elements is lessened, which cause the bending restoring force. Part of the bending energy is transformed into longitudinal energy because of the curvature. The strength of this transformation is determined by the transformation matrix $\mathbf{T}^{i,i+1}$. With a flat plate the amount of transformation is zero, with an angle of $90°$ the transformation is 100%.

Now, however, in addition to the bending, the longitudinal movement has to be considered. Here the coupling behaves in precisely the same way. The transformation matrix indicates how much of the distortion is passed on in the x plane of the ribs from one element to the next in the x direction. The rest of this longitudinal movement goes into the transverse movement in the z direction of the ribs.

The total transfer of energy between two elements can best be imagined using the example of a 90 degree angle. Here, a bending wave can no longer experience any restoring force. The upward movement in the z direction is, however, like a movement in the x direction for the bent part, i.e. a purely longitudinal one. At all bends – and thus also at all couplings from the back plate or the top plate to the ribs and vice versa – a longitudinal movement may be turned into a transverse one. This coupling of the two types of vibrations occurring in curved structures is not as radical as in the case of right angles, since here the curvature is less. But still there will be a coupling that must change the eigenvalues, because the two types of vibrations have different velocities of propagation. So if we look at the bending wave only, its eigenvalues must get higher. An impulse that displaces the ribs in the z direction at one particular point propagates itself along the ribs, not only in the z direction, but also in a "subterranean" manner in the longitudinal direction. As this longitudinal movement is faster than the transverse movement, and is transformed back into transverse movement for each curvature of the ribs, overall an increase in the velocity of the wave in the x direction of the ribs is to be expected. The eigenvalues of the plate will be higher than those of a similar, but flat, plate.

So, in both cases, (6.1) that of a plate curved by external pressure and (6.2) at of a plate curved owing to its geometry, without being under tension – the eigenvalues can be expected to increase, and this can also be measured experimentally. In the case of the ribs, this rise is not intended by the guitar-makers – unlike the rise found in the back plate. In the latter case the curvature is applied precisely because it is said to render the overall sound brighter and more brilliant. Brightening when a plate is under tension was also to be expected and the relevant calculations also confirmed this result. The increase in frequencies in the curved ribs was found, since the coupled longitudinal vibrations exist at higher frequencies, simply because of their higher wave velocity. But it is the transformation of these waves back into transverse waves that makes them audible, since only transverse vibrations can be radiated away from the instrument to any significant degree.

So the equations of the ribs for the bending waves are

$$\frac{\partial^2 u_z}{\partial t^2} = -\frac{EK^2}{\rho} \left(P_C \frac{\partial^6 u_z}{\partial x^6} + \frac{\partial^4 u_z}{\partial y^4} \right) - P_L \frac{E}{\rho} \frac{\partial^4 u_x}{\partial x^4} \tag{6.29}$$

and for the longitudinal movement in x-direction of the ribs

$$\frac{\partial^2 u_x}{\partial t^2} = -\frac{E}{\rho} \left(P_C \frac{\partial^4 u_x}{\partial x^4} \right) - P_L \frac{EK^2}{\rho} \frac{\partial^6 u_z}{\partial z^6} \ . \tag{6.30}$$

For reasons of simplicity, the shearing and the Poisson contraction are not included in the longitudinal movement here. Additionally we neglect the depiction of the LT-coupling of first order out of the same reasons. They have to be added to the terms of longitudinal displacement. The differential equation of the longitudinal movement in x-direction is enlarged, too, because of the curvature with a derivation of second order.

From the above discussion are are able to determine the values of P_C and P_L. First of all, they are the compensation for the geometrical curvature at the i-th node point, so

$$P_C, P_L \sim \cos(k(i)) \ , \tag{6.31}$$

secondly they describe the energy transport of the two kinds of waves along the x-component of the transformation matrix

$$P_C, \frac{1}{P_L} \sim \frac{Tx^{i,i+1} + Tx^{i-1,i}}{2} \ . \tag{6.32}$$

So it is

$$P_C = \cos(k(i)) \frac{Tx^{i,i+1} + Tx^{i-1,i}}{2} \tag{6.33}$$

and

$$P_L = \frac{2 \cos(k(i))}{Tx^{i,i+1} + Tx^{i-1,i}} \ . \tag{6.34}$$

These are the equations for the LT-coupling of second order. We now want to show some results for the ribs geometry.

The spectra of the ribs are shown for the curved and for the non-curved state in Figs. 6.8 and 6.9.

The first ten mode frequencies are shown below in Table 6.2 for the curved and the non-curved cases.

The curved eigenvalues in the lower range (1–4) are clearly higher than those of the non-curved ribs. In the range 5–10 there are three higher ones (5, 9 and 10), one identical one (5) and even two where the frequency of the non-curved ribs greater than that for the curved ribs (7 and 8). If we follow

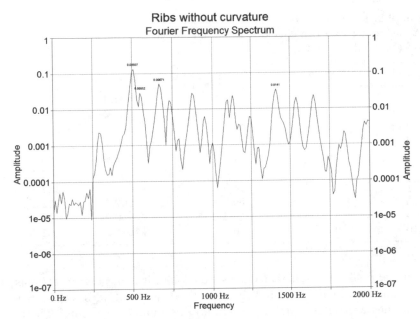

Fig. 6.8. Spectrum of the non-curved ribs (0–2000 Hz). The spectrum was calculated by applying an ideal Dirac impulse to a rib element. The spectrum was measured after the initial attack time

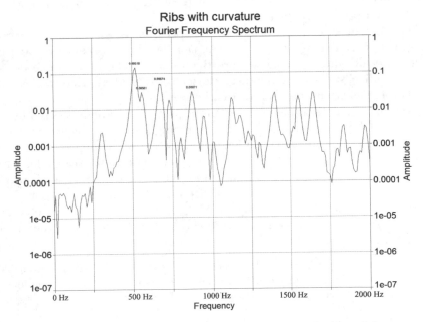

Fig. 6.9. Spectrum of the curved ribs (0–2000 Hz). For an evaluation of the spectrum see previous figure

Table 6.2. Frequency values (f) and amplitude values (A) for the ribs, with the ribs regarded either as a non-curved plate (f non-curved, A non-curved) or as a curved plate (f curved, A curved)

Mode	f Non-Curved	A Non-Curved	f Curved	A Curved
1	287.4	11.5	297.6	11.7
2	489.6	581.7	499.2	717.9
3	528	110.0	537.6	149.5
4	643.2	253.7	652.8	251.7
5	700.8	77.2	700.8	90.9
6	748.8	6.8	768	8.2
7	844.8	138.5	835.2	152.9
8	912	30.6	902.4	32.3
9	950.4	3.4	960.0	6.0
10	1046.4	80.8	1075.2	102.2

individual peaks in the higher ranges – although it is not always obvious how to allocate them here – we find that this tendency continues. On average the eigenvalues are raised, although there are individual exceptions. The reason must lie in the fact that only those eigenvalues can be raised, which run in the x direction of the ribs. The eigenvalues in the y direction do not experience any change, since the y direction is not curved. Combined mode frequencies only experience the effect weakly, so there are also some mode frequencies that remain practically unaffected.

The influence of the curvature on the spectrum is related to the average curvatures over the ribs as a whole. As can be seen, the numeric curvature values are all quite small. Therefore the values in the above table do not deviate radically from one another – in fact only slightly. Nevertheless, the increase in the eigenvalues can be clearly heard.

Another simulation was used to "knock" the ribs virtually again, but this time with damping was applied. As this damping is not dependent on velocity, but instead is dependent on the force (see section on damping), this means that the eigenvalues are not changed. But it was possible to create a knocking sound that could be played and heard. Here the audible impression allows one to conclude a slight raising of the frequencies of the curved ribs in comparison to non-curved ones. But a brightening of the sound can also be heard clearly. The higher frequency ranges appear to be slightly preferred in the curved version; this version of the ribs sounds brighter.[1]

---

[1] The same principle is known from the singing saw. Here, a saw is driven by a violin bow to sound in its eigenfrequencies. The melody is played by bending the saw and so increasing the fundamental pitch of the saw.

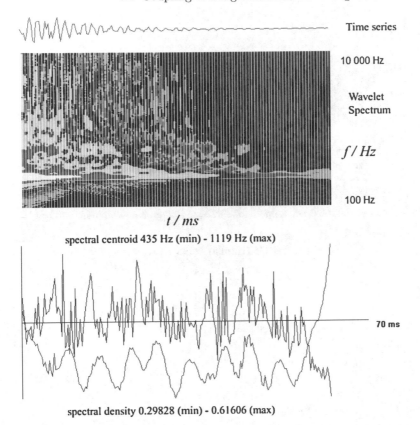

Fig. 6.10. Analysis of the "knocking sound", i.e. the application of a Dirac impulse to an element in the non-curved ribs. The 70 ms time series is included in a schematic form at the top, and underneath is its wavelet transformation, whose time axis runs parallel to the above time series. The vertical axis is the linear frequency, which goes from 100 Hz to 10 000 Hz. This is the range of the analysis used for calculating the spectral centroid and the spectral density

To check this, the spectral centroid was calculated (see Figs. 6.10 and 6.11)[2]. The overtones were weighted in relation to their position and amplitude. If the spectral centroid is high, the sound is bright; and if it is low, the sound is darker.

The development of the two centroids illustrates the tendency. In the curved case the curve of the centroid (blue) tends towards higher peaks – which can be observed over long stretches – in comparison to the centroid curve in the non-curved case, which is more subdued. And the deepest centroid

---

[2] The spectral centroid is scalar, and it weights the frequencies present in the sound in relation to their frequency and amplitude values. This produces a physical measure to add to the psycho-acoustic phenomenon of brightness.

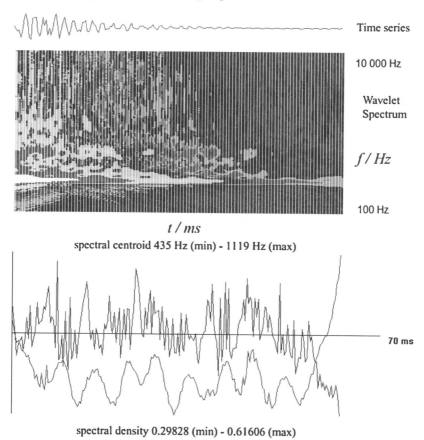

Time series

10 000 Hz

Wavelet
Spectrum

f / Hz

100 Hz

t / ms

spectral centroid 435 Hz (min) - 1119 Hz (max)

70 ms

spectral density 0.29828 (min) - 0.61606 (max)

Fig. 6.11. Analysis of the "knocking sound" on the curved ribs. For description see above Fig. 6.10

frequency of 377 Hz in the non-curved case has risen to 435 Hz for the curved ribs (the center lines of the plots are 746 Hz for the non-curved ribs and 777 Hz for the curved ribs).

The spectral density speaks the same language. It is defined as sound entropy, that is the density of the spectrum at any point in time. Here there are two extremes. If only one sine wave tone is present, that is, if only one peak exists in the spectrum, there is maximum order. The spectral density is zero. If there is silence, or white noise, all frequencies are present in equal amounts. The spectral density is at a maximum, equal to one. Here the density for the non-curved ribs moves in the range 0.31 to 0.68, and for the curved ribs in the range from 0.30 to 0.62. The curved case is thus somewhat less dense that the non-curved one. This is what would be expected. The slight shifts in frequency are of no importance for the density. But if all the peaks are raised a little, as in a brightening of the sound, then the brighter peaks are more different from smaller peaks and also from background noise. It would be expected that the sound comes over as less dense, i.e. somewhat brighter. This is indeed the case and it was possible to calculate it here.

Another point to observe is the falls in the centroid at the temporal endpoints of the plot and the dramatic climbs in the spectral densities at these points. The centroid must become smaller, the fewer overtones are present. In the case of percussive sounds, it can be expected that while the overtones die away quickly, this also happens to the fundamental. This can also be seen in the wavelet plot. Towards the end of the 70 ms shown here, however, the fundamental is still sounding, while the overtone structure has collapsed. The centroid then falls to the value of the fundamental, which in this case is around 290 Hz, i.e. below the scale shown.

The rise in the spectral density towards the end is analogous to this. Towards the end, the fundamental has virtually died away completely. It almost disappears in the noise. Noise, however, has a spectral density of one. The plot must move towards one.

Overall, one should consider that the change in sound shown here is small. It is not minimal, but it is not very large, either. This means that it is moving within the range that is musically interesting. If an extreme change in the sound were to occur, the fundamental character of the instrument would also alter. The guitar would turn into a different, but related, instrument. It might start to sound something like a harpsichord or a zither. But this is never what is intended. All the changes are intended to be audible, and the instrument should still sound like a "guitar". If the observed changes were extreme, one would have had to exercise caution as to whether the changes were acceptable or not. But here we find precisely what musicians and instrument-makers are interested in, namely a slight nuance in the sound, intended to influence the character of the instrument. What is required here is not more, but not less than this either.

6.2 Time-Stepping Method

The most common time stepping method, both for the Method of Finite Elements (FEM) and for the Finite Difference Method (FDM) used in this work, is the Newmark procedure (Newmark 1959)

$$\mathbf{u}^{t+\triangle t} = \mathbf{u} + \mathbf{v}\triangle t + \frac{\triangle t^2}{2}((1-2\beta)\mathbf{a} + 2\beta\mathbf{a}^{t+\triangle t}) \qquad (6.35)$$

Starting from a displacement $\mathbf{u}$ and a velocity $\mathbf{v}$ at a time point t, it calculates the displacement and velocity for the next point in time $t + \triangle t$ with three terms.

$$\mathbf{u}^t \quad \text{Term (1)} \qquad (6.36)$$

Term (6.1) is merely the displacement from the previous point in time t.

$$\mathbf{v}\triangle t \quad \text{Term (2)} \tag{6.37}$$

Term (6.2) is the change in displacement owing to the velocity v, which the body has for a period of $\triangle t$ seconds.

$$\frac{\triangle t^2}{2}((1 - 2\beta)\mathbf{a} + 2\beta\mathbf{a}^{t+\triangle t}) \quad \text{Term (3)} \tag{6.38}$$

Term (6.3) denotes the change in displacement owing to the acceleration $\mathbf{a}$, which the body experiences during the period $\triangle t$.

The different variants of the Newmark procedure (giving the Newark family its name) arise from the value for β. If $\beta = 0$, term (6.3) becomes

$$\frac{\triangle t^2}{2}\mathbf{a}\,. \tag{6.39}$$

The displacement from the next point in time $t + 2 * \triangle t$ is thus the old displacement plus its change owing to the velocity and acceleration of the body at the previous point in time, over the time interval $t + \triangle t$.

$$\mathbf{u}^{t+\triangle t} = \mathbf{u} + \mathbf{v}\triangle t + \frac{\triangle t^2}{2}\mathbf{a}\,. \tag{6.40}$$

$$\mathbf{v}^{t+\triangle t} = \mathbf{v} + \triangle t((1 - \gamma)\mathbf{a} + \gamma\mathbf{a}^{t+\triangle t}) \tag{6.41}$$

With the central difference method the weighting term γ becomes $\gamma = 0.5$. In the version of the central difference method used here, $\gamma = 0$. Therefore also the velocities at time $t + \triangle t$ are calculated out of the velocities at time point t and we need no initial knowledge of the accelerations at the following time point $t + \triangle t$. The change in velocity thus becomes

$$\mathbf{v}^{t+\triangle t} = \mathbf{v} + \mathbf{a}\triangle t\,. \tag{6.42}$$

The acceleration at the next point in time depends both on the geometric conditions, i.e. how the element stands in relation to its neighboring elements, and on the couplings that it experiences. After calculation of all these influences, we can calculate the change in displacement for every point in time $t + n * \triangle t$ with $n = 1, 2, 3, \ldots$ Therefore $\triangle t$ is the time interval, which in our musical case is the reciprocal of the sampling frequency S_f, where $S_f = 1/\triangle t$. But there is no reason not to change $\triangle t$ during the calculation in any way we choose. For time periods in which the system moves rapidly, wherever high frequencies occur $\triangle t$ must be small. But when the guitar after the string pluck and an initial transient phase enters the so-called steady-state, the higher frequencies have already died away. Then the value for $\triangle t$ can be enlarged, thus saving time spent on calculation. We will come to a discussion of the optimum values for $\triangle t$ below.

Setting β and γ to zero has the following reason. With values $\beta \neq 0$ and $\gamma \neq 0$ in the calculation there appear the difficulty, which is avoided if these values are set to zero like used here. Because then we would have to know **a** not only at the time point t, but also at the time point $t + \triangle t$ to calculate **v** and **u** to the time point $t + \triangle t$. For calculating the next step, therefore, at least the acceleration of this next step must be known. In practice, the accelerations, and thus the overall structure of the body, must be assumed at two successive points in time. This can often be very difficult, if not impossible, for complicated bodies such as the guitar, if one does not want to suppose two arbitrary displacements, because the complicated geometry would make the necessary and realistic guess of **a** at all node points very difficult, also as trivial accelerations (i.e. **a** = 0 for t and $t + \triangle t$) are impossible, if the real situation of a guitar tone shall be simulated.

The advantage of these procedures with values of $\beta \neq 0$ and $\gamma \neq 0$ lie in the fact that the time-step can be considerably enlarged without compromising the stability of the procedure. Every procedure has its own critical S_f. Hughes (Hughes 1987) gives the stability criteria for the best-known procedures of the Newmark family after (Goudreau and Taylor 1972; Krieg and Key 1973) and (Hilber 1976).

The formula of the critical sampling frequency in relation to the still representable frequency determines $\triangle t$.

$$S_\omega^{\text{crit}} \geq \triangle t \, \omega \tag{6.43}$$

So if S_f^{crit} increases, we require a larger $\triangle t$ to realize the same angular frequency ω. As larger the value of S_f^{crit}, the larger is the range of possible values for $\triangle t$. Let us examine the case of a system to which the Nyquist frequency theory applies. Here every sine wave frequency must be represented by at least two points in time. If we write (6.43) using the frequency f instead of the angular frequency ω, to put the results into Table 6.3, we get

$$S_f^{\text{crit}} \geq \triangle t \, f \,, \tag{6.44}$$

and for this study it holds $S_f^{\text{crit}} = 2$, because of the Nyquist condition. If we want to represent a frequency $f = 1000\,\text{Hz}$, then $\triangle t \leq \frac{1}{2000}$. For every higher

Table 6.3. Comparison of various time-stepping methods. They can be of either an implicit or an explicit type. The values of the Newmark procedure are given for the calculation of the displacement and velocity together with the critical sampling frequencies. (Taken from Hughes 1987 p. 493)

Method	Type	β	γ	Stability Cirterion
Average acceleration	implicit	$\frac{1}{4}$	$\frac{1}{2}$	Unconditional
Linear acceleration	implicit	$\frac{1}{6}$	$\frac{1}{2}$	$S_f^{\text{crit}} = 3.464$
Fox-Goodwin	implicit	$\frac{1}{12}$	$\frac{1}{2}$	$S_f^{\text{crit}} = 2.449$
Central differences	explicit	0	$\frac{1}{2}$	$S_f^{\text{crit}} = 2$

Table 6.4. Comparison of various time-stepping methods in relation to the required time-steps $\triangle t$ per period T. (Taken from Hughes 1987 pp. 541f)

Method	$\triangle t/T$
central differences	0.000318
trapezoid rule	0.0125
damped Newmark method	0.003
α-method	0.010
Wilson-θ-method	0.008
Houbolt method	0.006
Park method	0.008

value of $\triangle t$ this representation of the sine wave no longer applies. In complex iterative systems S_f^{crit} is often dependent on the system. In such cases $\triangle t$ has to fit in with both this and the desired maximum frequency to be represented (see Table 6.4).

In (Hughes 1987) p. 541, the required values $\triangle t/T$, i.e. the values for one time period, from various methods are compared.

The trapezium rule is the above-mentioned average acceleration ($\beta = 1/4$, $\gamma = 1/2$). The enlargement of the size of the steps can incur an unwanted increase in amplitude values at the high frequencies. To avoid this, the damped Newmark method ($\beta = 0.3025$, $\gamma = 0.6$) has been developed. Both the α method (Hilber Hughes and Taylor 1977) and the Wilson-θ method attempt to combine a high degree of stability with low overtone reinforcement. These methods are multi-step algorithms. They try to extract the movement from k points in time. The Houbolt method weights these points in time. The Park method uses $k = 3$.

Here in musical acoustics, the important thing is not necessarily a procedure that allows large time-steps. Larger time-steps are generally needed when the temporal development itself is not of interest, but instead an end value upon which the system is to converge. One example would be an iron support that is heated from underneath and passes on some of this heat upwards. After a transient period of time, the system converges on a certain temperature. The details of the way in which the system arrives at this temperature are not of any interest. However, in the simulation of musical instruments it is precisely the transient that is important. The smaller the possible resolution of the temporal development, the smaller $\triangle t$ is; and the larger S_f, is, the more exact a later analysis can be using signal processing. Apart from this, the exact spatial development of initial transients in the instrument have decisive effects on its audible properties (Reuter 1995; Beurmann and Schneider 2003). So, if one wants to obtain detailed findings, sampling rates of 300 kHz–500 kHz would be desirable. For the stationary state a sampling rate from 96 kHz is often sufficient. At this point we should remind ourselves that the sampling frequency of a CD is 44.1 kHz i, although this is often exchanged for 96 kHz in the studio, and that DVD players operate with a sampling rate of 192 kHz.

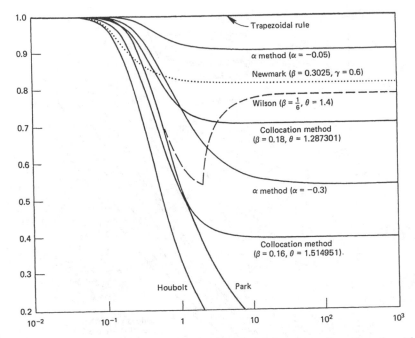

Fig. 6.12. Damping plot for high frequencies for various time iteration algorithms. The damping ϱ with $\varrho^n \le e^{-n(1-\varrho)}$ is measured in radians of the relationship $\Delta t/T$. (Taken from Hughes 1987 p. 533)

The type of central differences used here, which only requires the assumption of the acceleration of a time-step to calculate the following state of the system, likewise possess a critical sampling frequency of $S_f^{\text{crit}} = 2$. So theoretically, the smallest frequency that can be represented is half the sampling frequency, the Nyquist frequency. Unlike the situation with digital-analog and analog-digital converters, this is precisely the case here. For these converters the Dirac impulse in the sampling process must be infinitesimal in order to fulfill the above condition – which cannot be realized. Here in the simulation of musical instruments the impulse is in fact infinitesimally short, so the Nyquist frequency does actually mark the boundary. This can very easily be seen in the spectra of the sounds created. Up to a certain frequency the spectrum corresponds to that of the sound, but then falls away suddenly at the Nyquist frequency practically to zero (see also Figs 6.12 and 6.13).

6.3 Energy Conservation

In time dependent problems, the energy conservation in not automatically implemented in the method, but just given implicitly. If the displacement of the system is carried out from one point in time to the next according to one

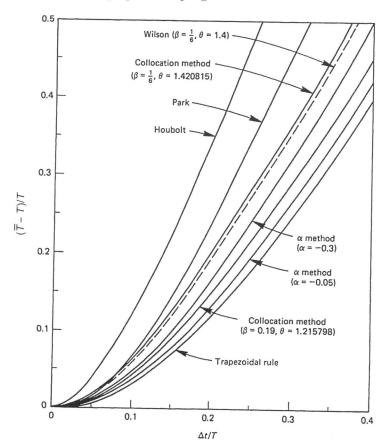

Fig. 6.13. Relative errors per time period dependent on the duration of the period T for various methods. The method applied in this monograph, with a range of 300 kHz to 500 kHz, would be $\Delta t/T$ from 1/300–1/500 or a frequency of 1 kHz, that is on the far left boundary of the plot, and for 20 kHz at 0.067–0.04 even with very low values. (Taken from Hughes 1987 p. 535)

of the procedures mentioned above, theoretically this guarantees conservation of energy. The laws of mechanics apply. For this reason the energy in the system must remain constant. But whether this is really the case, owing to the discrete formulation, first has to be clarified (Wriggers 2001) S. 204ff (Haile 1997) S. 163ff.

There are several methods for minimizing errors when using FDM. One of the procedures is a general procedure for correcting the calculation. Once the displacement and velocity to the next time-step have been calculated, this is corrected by applying general laws, such as the general solution for bending vibrations, or Eulers method, in order to turn the calculated value into the actual one.

Another procedure is Gears processing correction method (Gear 1971). He achieves a high level of accuracy by calculating the displacements and accelerations to the next step with the changes up to the fifth derivative. The displacement

$$\mathbf{u}^{t+\triangle t} = \mathbf{u} + \frac{d\mathbf{u}}{dt}\triangle t + \frac{d^2\mathbf{u}}{dt^2}\frac{\triangle t^2}{2!} + \frac{d^3\mathbf{u}}{dt^3}\frac{\triangle t^3}{3!} + \frac{d^4\mathbf{u}}{dt^4}\frac{\triangle t^4}{4!} + \frac{d^5\mathbf{u}}{dt^5}\frac{\triangle t^5}{5!} \ , \quad (6.45)$$

the velocity

$$\frac{d\mathbf{u}^{t+\triangle t}}{dt} = \frac{d\mathbf{u}}{dt} + \frac{d^2\mathbf{u}}{dt^2}\triangle t + \frac{d^3\mathbf{u}}{dt^3}\frac{\triangle t^2}{2!} + \frac{d^4\mathbf{u}}{dt^4}\frac{\triangle t^3}{3!} + \frac{d^5\mathbf{u}}{dt^5}\frac{\triangle t^4}{4!} \quad (6.46)$$

and the acceleration

$$\frac{d^2\mathbf{u}^{t+\triangle t}}{dt^2} = \frac{d^2\mathbf{u}}{dt^2} + \frac{d^3\mathbf{u}}{dt^3}\triangle t + \frac{d^4\mathbf{u}}{dt^4}\frac{\triangle t^2}{2!} + \frac{d^5\mathbf{u}}{dt^5}\frac{\triangle t^3}{3!} \quad (6.47)$$

are calculated by the higher derivatives.

$$\frac{d^3\mathbf{u}^{t+\triangle t}}{dt^3} = \frac{d^3\mathbf{u}}{dt^3} + \frac{d^4\mathbf{u}}{dt^4}\triangle t + \frac{d^5\mathbf{u}}{dt^5}\frac{\triangle t^2}{2!} \quad (6.48)$$

$$\frac{d^4\mathbf{u}^{t+\triangle t}}{dt^4} = \frac{d^4\mathbf{u}}{dt^4} + \frac{d^5\mathbf{u}}{dt^5}\triangle t \quad (6.49)$$

$$\frac{d^5\mathbf{u}^{t+\triangle t}}{dt^5} = \frac{d^5\mathbf{u}}{dt^5} \quad (6.50)$$

All the above equations refer to the displacement and its derivatives from the time-step t, if there is no higher power present.

In the present work, first the force between two elements is calculated. This is done using differential equations of the second order. The calculation corresponds to the relevant system and is carried out over the energy. An acceleration can be calculated from the force. The difference between the acceleration calculated from the force and that calculated from the system of equations is

$$\frac{d^2\triangle\,\mathbf{u}}{dt^2} = \frac{d^2\mathbf{u}^{\text{Kraft}}}{dt^2} - \frac{d^2\mathbf{u}^{\text{calculation}}}{dt^2} \ . \quad (6.51)$$

This corrective term is now used to correct the displacement and its derivatives:

$$\mathbf{u} = \mathbf{u}^{\text{calculation}} + \frac{\alpha_0}{2!}\frac{d^2\triangle\mathbf{u}}{dt^2}\triangle t^2 \quad (6.52)$$

$$\frac{d\mathbf{u}}{dt}\triangle t = \frac{d\mathbf{u}^{\text{calculation}}}{dt}\triangle t + \frac{\alpha_1}{2!}\frac{d\triangle\mathbf{u}}{dt^2}\triangle t^2 \quad (6.53)$$

$$\frac{1}{2!}\frac{d^2\mathbf{u}}{dt^2}\triangle t^2 = \frac{1}{2!}\frac{d^2\mathbf{u}^{\text{calculation}}}{dt^2}\triangle t^2 + \frac{\alpha_2}{2!}\frac{d^2\triangle\mathbf{u}}{dt^2}\triangle t^2 \quad (6.54)$$

$$\vdots \quad (6.55)$$

Table 6.5. Values for α in Gears processing correction method algorithm for the levels of accuracy $q = 3$, $q = 4$ and $q = 5$

α	$q = 3$	$q = 4$	$q = 5$
α_0	$\frac{1}{6}$	$\frac{19}{120}$	$\frac{3}{16}$
α_1	$\frac{5}{6}$	$\frac{3}{4}$	$\frac{251}{360}$
α_2	1	1	1
α_3	$\frac{1}{3}$	$\frac{1}{2}$	$\frac{11}{18}$
α_4	$-$	$\frac{1}{12}$	$\frac{1}{6}$
α_5	$-$	$-$	$\frac{1}{60}$

The values of α are given for several degrees of precision (see Table 6.5).

The values of a are given for various levels of accuracy. Most of these correction algorithms are based on second order differential equations. It is true that they can be applied to higher orders, but in practice in the case of a curved structure it is very complicated to describe the procedures with all the couplings and the differential equations of the sixth order described above. So we come up against the question of whether energy is conserved in the procedures used here and with the small time steps, or whether it is dissipated, or fluctuates.

The total energy of the system of a plate fixed at all boundaries equals the sum of the potential and kinetic energy

$$E = T + K \tag{6.56}$$

where T is the kinetic energy and K he potential energy. In the case of the plate E is given by the equation

$$E = \frac{1}{2}mv^2 + \frac{1}{2}Ds^2 \tag{6.57}$$

for a longitudinal vibration and a differential equation of the second order, with m being the mass and E the proportional constant between stress and strain, σ the stress and ϵ the strain with

$$\sigma = E\epsilon \tag{6.58}$$

E can be interpreted as spring stiffness if s is the displacement of the node points of the geometry from their equilibrium position.

The diagram gives the total energy of a plate over the calculated iterations. The plate is a palisander plate measuring 2×2 meters with a thickness of 3 cm, and in the middle it has been displaced longitudinally in the x direction, i.e. one element has been displaced by 1 cm from its position at rest. Although this is not possible in practice, it corresponds to a Dirac impulse and can be applied here. The energy is independent of the point at which the displacement occurs. The displacement in the x/y direction was selected (here without

Energy

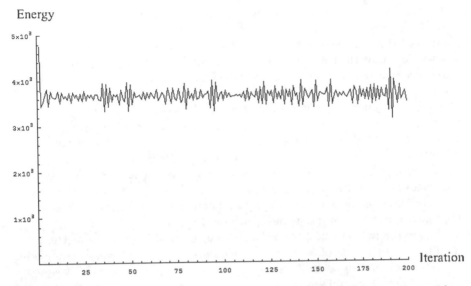

Fig. 6.14. Plot of the total energy over the iterations for a 2×2 meter plate with a thickness of 3 cm. The average energy of $E_D = 3.67e8$ remains constant. Fluctuations take place, which correspond, for example, to thermal fluctuations, noise, etc

coupling with the transverse movement in the z direction) to enable use of the above definition of the total energy. This applies to the energy of a second order differential equation. In the case of bending waves the energy also resides in the curvature. The results of these calculations do not differ significantly from the results presented here.

At first the unusual stimulation causes a certain energy value, which is never repeated. The following energy series remains constant on average, with $ED = 3.67e8$, while the fluctuations are considerably smaller (see Fig. 6.14). The behavior is typical of such iteration processes. In the simulation of molecular dynamics these fluctuations are treated as random thermal fluctuations.

Here this would correspond to something like random fluctuations arising from "mechanical thermal activity", that is, random fluctuations. We have to keep in mind, that the laws of classical mechanics describe the global behavior of the system without knowledge of all the small subsystems, molecules ect. A good example of this is friction. It is hard to define the cause of friction (Urbakh et al. 2004). If we were to put two metal plates with perfectly smooth surfaces one on top of the other, there would be no physical differentiation possible between the layers of atoms resting on top of one another and those within the plates. The plates would be fused together and the "friction" would equal the force necessary to rip a plate down the middle. Friction thus is caused by a surface that is rough or dirty. This applies to external friction just as it does to the processes in the plate itself. These numerous small processes, which are nonlinear by nature, lead to a fluctuation in energy.

Naturally, this fluctuation is not included in the calculation, but it can be read off here as a result. The fluctuation is caused by two sources of error. First, an error can be expected since the elements of the plate have a finite length. Secondly, the time interval of the iteration is finite. The effects of these two sources of error become stronger when the spatial and temporal intervals are larger. But it remains astounding that the average total energy does not change. Even after many iterations it stays the same. Interestingly, however, the cause of these fluctuations can be traced back to a nonlinearity. For every element C8 is stable. However, wherever the elements meet one another, a curvature is necessary. The elements are assumed to be linear. They could also be assumed to be polynomials with terms of a higher order. But even then the interfaces between them would not be smooth. But the assumption is always that the time and space intervals are so small that a stable development can be approximated, even if it is never achieved exactly.

Interestingly, the cause for this fluctuation in the iteration method is a nonlinearity, here one of an unsteady geometry. The polynomials connecting the node points is in itself C^∞ steady. But at the node point itself it is necessarily a crack. This crack is not caused by the fact, that the element functions used here are linear. The element functions could also be assumed as higher order polynomials. But even then the structure at the node points would not be steady. This holds in general for all discrete methods, Finite-Difference methods, Finite-Element methods or similar algorithms. The assumption always is, that the temporal and spacial intervals are so small, that a steady curve can at least be approximated, even if it is never reached exactly. This keeps the fluctuations described above in a small range.

6.4 Differences in Spacial and Temporal Representation Between Discrete and Continues Mechanics

Continues and discrete mechanics differ in their notions of space and time in some points, which shall be discussed here briefly. Some of these differences are essential, concerning the fundamentals of the respective methods, some are by-products, which have to be coped with in the algorithms.

The most important difference between the plate as a differential equation and as a discrete formulation is that of time symmetry. In the differential equation time symmetry holds. In the discrete mechanical case, time symmetry is broken.

The Newtonian idea of time as continues aches, which mathematically can be treated as a spacial axis leads to the possibility of time reversal. If time is reversed (by changing t to $-t$), all time dependent physical parameters have to change too, like speed, acceleration etc. Reversal is understood in a way, that the absolute value (i.e. $|v|$ with velocity) is conserved and just the sign is changed. If a ball have a velocity $\mathbf{v}$ in time t, then after reversal of time at

$t_u = -t$ it has the velocity $\mathbf{v}_n = -\mathbf{v}$. So if we let the film go backward, the ball takes the same trajectory back.

This means with the plate, that the movement is just reversed with time reverse. So if we struck the plate and let it move of say 10 ms, we would arrive at exactly the same starting displacements, if we reverse time and therefore velocity at $t = 10$ ms and move back for again $\triangle t = -10$ ms.

The Figs. 6.15 and 6.16 of the top plate show the plates reaction to a plucking of a guitar string from $t = 0$ ms up to $t = 0.666663$ ms in twenty steps of $\triangle t = 0.034722$ ms the expected development. If we apply the time reverse at the tenth time step, the vibrations reverse as shown below.

The movement does not lead back again. At first sight, the vibration seem to calm down again, but after a close look one realizes small deviations from the zero state over the whole plate. These small deviations one the one hand side have the cause in the truncation error of spacial and temporal finite steps. But secondly, this deviations is caused by the singularity of the situation at the beginning of the vibration. If the time reverse operation would be applied to a standing wave, this time reverse would still hold. But here, a rest error is left behind, which is of systematical nature.

For simplicity we assume a single node point to be displaced similar to the case of knocking at the top plate. During the next time step, this node point tears all neighboring points a little bit in the direction of the initially displaced node point. This initial displaced node point itself looses a bit in displacement and move down a bit. The velocities of the neighbor elements are now positive, the velocity of the middle element is negative. If we now assume a time reversal which reverses the velocities, then the middle node point has a positive velocity and again is its way up. The neighbor elements now have a negative velocity and are on their way down again.

But now, at each point in time, the influence of one node point on all other neighboring node points has to be taken into consideration. At zero time all elements have been accelerated around the middle element. But at the time point after one iteration, if time is reversed, now also the node points next the the neighbor node points of the middle element are influenced. So also they are moved. The middle node point indeed is driven its way up again, but never reaches its initial spacial height. A little bit of energy now flows into the neighbor elements; a small wave leaves the node points to both sides. The initial state is never reached again. Time symmetry does no longer hold.

This problem can not be diminished by giving each node point a larger circle of possibly influencing node points. This circle will ever find an end, even if this end is the plates boundary. At least then a deviation will appear and the initial state is not reached.

So here we have a second difference between continues and discrete mechanics. The continues mechanics assumes, that each point is instantaneous entangled with all other points of the geometry. If we throw a stone into water, the wave will spread with the velocity of wave in water. But to fulfill the differential equation which leads to this wave, we have to assume, that at timepoint

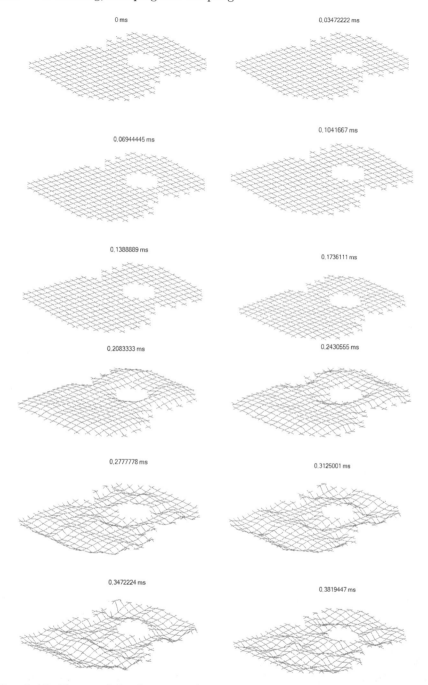

Fig. 6.15. Temporal development of a guitar top plate from time point $t = 0$ of plucking a string and further time steps each $\triangle t = 0.034722$ ms apart. The calculation has been visualized by the MTMS software. Here the visualization of the top plate was used and "snap shots" were taken at the mentioned points in time

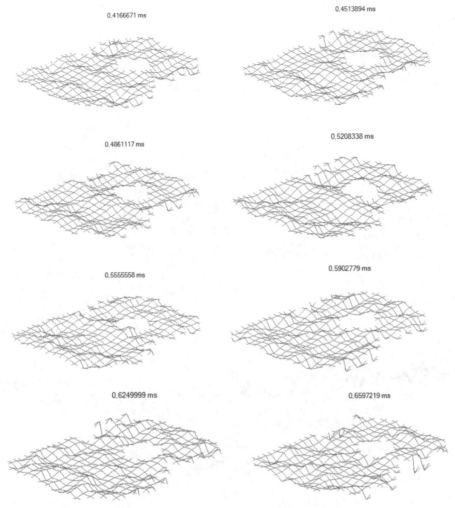

0.4166671 ms

0.4513894 ms

0.4861117 ms

0.5208338 ms

0.5555558 ms

0.5902779 ms

0.6249999 ms

0.6597219 ms

Fig. 6.15 (*continued*)

zero, also the most distant corner of the water basin gets an impact. Strictly speaking this not only holds for the water basin, because the wave will break at its shore, but for the whole universe. If we leave out relativistic calculations from our discussion, then continuum mechanics claims instantaneous impact.

This does not hold for discrete mechanics. As we have seen, the influence of the node points is restricted to their surroundings, according to the wave speed in the medium. If the time step is short enough, so we are largely in the range of the wave speed in the considered medium, then the calculation *in case time is now reversed* will represent the measurements.

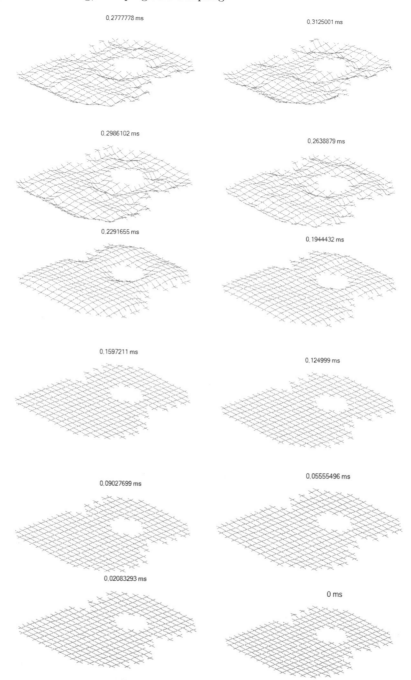

Fig. 6.16. Temporal development of a guitar top plate from time point $t =$ 0.333333 ms after plucking a string and after reverse of time, here the reverse of the node point velocities. The development does not lead back. For realization of the plots, in the MTMS software at the mentioned point in time the time reverse button was pressed

So we see, that in continuum mechanics, time symmetry can just be claimed by assuming an idea of space and spacial impact, which is contraintuitive to nature. There is no reason to assume, that each point of a geometry is instantaneous related with all other points. Discrete mechanics on the other hand gives up this instantaneous influence – first of course due to purely practical reasons of calculation cost – but without mistake in the calculation. But it *does* break time symmetry.

The problem is similar to molecular dynamics (Finite-Difference Methods are indeed often used in simulations of molecular dynamics). The second law of thermodynamics describes the increase of entropy (mostly H), of disorder of a system, where no energy is applied to with increasing time. The first equations of L. Boltzmann in 1872 of the problem seemed to be able to solve the entropy problem as molecular system. Just later is showed up, that this would apply a fundamental time symmetry breading. All further tries to find a solution here failed (see (Wolfram 2002) p. 1019f). Just the information theory of Shannon could solve the problem, unfortunately no longer on a molecular level. But as we are on a kind of spacial distribution level with the Finite-Difference Methods here rather than at a statistical representation, we have to deal with that problem here, too.

At this point, we have to keep in mind, that time symmetry is a pure hypotheses, not a proved theory, if we assume, that physically the proving experiment makes a theory of an hypotheses. As time reversal never have been experienced, time symmetry sill is a hypotheses.

Additionally, the formal treatment of space-time with a vector of four components is not to be set equal, that time is a fourth dimension. Time in this formulation still keeps its properties as time, i.e. just moving into positive direction and just movements with the speed of light let time stand still, but never drive back. Space-time just explains the dependencies between space and time and not assumes the sameness of the two phaenomenae.

Speaking more intuitive, we can make the analogy with movies, which are presented backwards. Here, we reverse all velocities and therefore all movements. But, different from our case of the plate, it is not time we reverse. Time itself still goes into its positive direction. Strictly speaking, we just reversed the velocities. So we experience time as going forward. We just defined the direction as positive.

Another possibility of treating the time symmetry breaking is to look at it in terms of entropy.

The figures show the development of entropy of a plate. The entropy S is calculated by the Kolmogorov equation

$$S = \frac{1}{\ln(nx \; ny)} \sum_{ix=1}^{nx} \sum_{iy=1}^{ny} u_z(ix, iy) \; \ln(u_z(ix, iy)) \,, \qquad (6.59)$$

when the plate has nx times ny elements, which are displaced in z-direction by $u_z(ix, iy)$. As the displacements of the plate are normalized, its maximum

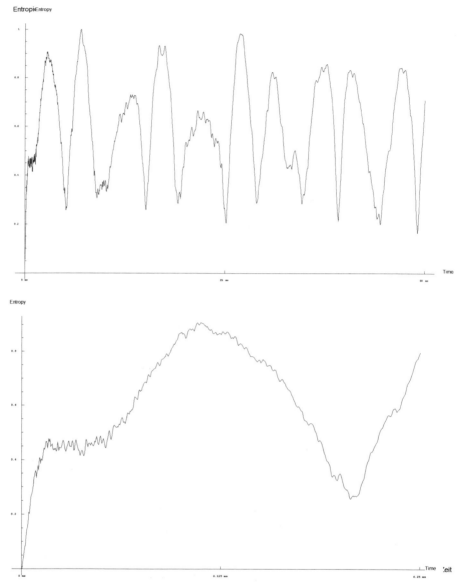

Fig. 6.17. Development of entropy of a plate, which is struck in its middle. At the beginning the system is in complete order with $E = 0$ (*figure below*). Then the spreading of the wave enlarges and so the degree of order in the plate decreases. The strong fluctuations (*figure above*) are assigned to the fundamental mode. At each zero crossing the plate is nearly flat and so the entropy is much larger. At the end of the entropy series shown above, the valleys get deeper, what is caused by the damping of the higher modes and so causes higher degrees of order

degree of order is reached, when just one element is displaced and all other elements have zero displacement. Then $E = 0$. If the plate has a random distribution of displacement values, then $E = 1$.

In Fig. 6.17 it can be seen, that the plate, which is struck at the beginning and so has just one displaced element at time point $t = 0$ has an entropy of $E = 0$. Then E increases and starts to oscillate. This oscillation is caused by the fundamental mode of vibration. The larger the displacement of the elements, the smaller the entropy. If the fundamental mode passes its zero crossing and so is zero in amplitude, only the overtones with their amplitudes being much smaller than that of the fundamental contribute to the entropy. E increases, because no definite structures can be seen any more.

Furthermore the tendency can be seen, that the movement tends towards zero with increasing time. Here the influence of damping shows up quite clearly, which first damps the higher overtones. Therefore during the zero crossings of the fundamental, the distributions of displacements decreases and E is getting smaller.

Keeping the reasoning above in mind, we clearly see, that the entropy value of zero at the beginning is never reached again. There is no reversability of events. To establish order again, which has reigned at the start, we had to get energy into the system again. This would also be the case if the plate were not damped. Because to reach the initial conditions again, all elements except of one must be zero again. This is a state of the system, it can never reach again by itself. This nonreversability of events can serve as a second explanation of the problem discussed above, which breaks the time symmetry.

6.5 Results for the Plate

This section will provide a comparison of the results from the Finite Difference methods (FDM) and and Finite Element methods (FEM) for a plate that is held fast. In the case of a rectangular plate that is fixed on all sides the following types of vibrations occur.

Figure 6.18 shows the first eight fundamental modes of a square plate. They are labeled with the number of node lines, which arise across or along the plate, written in brackets. $(0, 0)$ is the fundamental mode, shown top left in the diagram. This mode has no node lines, which means that the whole plate vibrates upwards and downwards as a single unit. The next higher frequency comes about when a node arises either longitudinally or transversely, labeled here $(1, 0)$ and $(0, 1)$. As the plate is totally symmetrical, its frequencies are the same, and the ratio to the fundamental mode $(0, 0)$ corresponds to $f2/f1 = 2.04$. In the frequency spectrum these two types of vibrations cannot be distinguished from each other. The next higher frequency comes

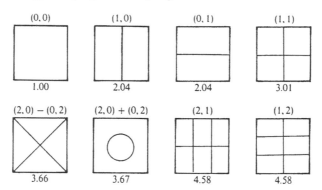

Fig. 6.18. The first eight modes of vibration of a plate fixed at the boundaries. The mode node number are given in brackets. The values under the plates are the frequency relations to the fundamental mode (0,0). From: (Fletcher and Rossing 2000) S. 87

about through the $(1, 1)$ mode, which has one node in the transverse direction and one node in the longitudinal direction. Its ratio to the fundamental frequency is 3.01. The two following modes $(2, 0)(0, 2)$ and $(2, 0) + (0, 2)$ are called degenerate. If two node lines occur longitudinally and transversely, they have the same frequency. Now there are two possible cases. The simultaneous vibrations can have two phase states. If the two areas that are on the outside vibrate in phase with each other, the whole edge also vibrates up and down in phase. We have a nodal circle in the center at $(2, 0) + (0, 2)$. In the other case the external sections move upwards longitudinally, while they move downwards transversely. This out-of-phase motion creates a nodal pattern in the form of a cross in the $(2, 0)(0, 2)$ mode. Since these types of vibrations influence one another slightly, the frequencies arising likewise vary slightly. Their ratios to the fundamental frequency are 3.66 and 3.67. The last modes are $(2, 1)$ and $(1, 2)$, shown with a ratio of 4.58 to the fundamental mode. These two modes cannot be distinguished from each other in the spectrum, either.

The values for a plate of 40 cm by 40 cm with a thickness of 3 mm, which is completely fixed along all its edges, i.e. the boundary conditions $\mathbf{R}_u = 0$ are satisfied, are calculated with the aid of the Finite-Difference Method and the Finite-Element Method. The frequency ratios can now be compared with the theoretical values.

The Tables 6.6 and 6.7 below provide an overview of the results from the calculations for FDM and FEM and allow them to be compared with theoretical values.

The values for the Finite Differences are very accurate, for the mode $(1, 0)$, with 2.06 instead of 2.08, and the mode $(2, 1)$, with 4.60 instead of 4.58. There are deviations for $(1, 1) : 2.87$ instead of 3.01, and for the degenerate mode $(2, 0)(0, 2)$, with 3.95 instead of 3.66. Since one value is somewhat too low and the other somewhat too high, we cannot speak of a general divergence of the spectrum, although the next higher value agrees very well.

Table 6.6. Frequencies of a plate 40 cm by 40 cm and 3 mm thick, calculated using FDM and FEM. The FEM values were calculated with various dampings and numbers of elements. MD: mass damping, SD: shearing damping, SF: shearing factor

Mode	FDM Clammed	FEM Clammed 3032 Elements	FEM Simply Supported 758 Elements	FEM Simply Supported 3032 Elements	FEM Simply Supported 3032 Elemente
		$MD = 1$ $SD = 0$ $SF = 1.2$	$MD = 1$ $SD = 0$ $SF = 1.2$	$MD = 1$ $SD = 0$ $SF = 1.2$	$MD = 1$ $SD = 0.01$ $SF = 1.2$
1	82.031	81.4	61.0	60.8	60.9
2	158.7	166.2	153.3	152.3	152.3
3	235.8	245.2	246.4	243.8	243.8
4	323.7	298.5	309.2	305.2	305.2
5	377.3	374.5	403.8	397.2	397.0

The results from FEM agree very well for the fixed plate. The values for the supported plate have been added to show that markedly different values occur here. In the case of the fixed plate, the boundary is not allowed to rise, quite apart from the fact that it cannot be displaced. The differences in the frequencies can be seen quite clearly.

The images for FDM and FEM of the plate frequencies are shown in Figs. 6.19 and 6.20. The modes must be actuated to produce the FDM diagrams. The actuations occur in succession with the frequencies measured for this plate. The actuation must occur at those locations that exhibit a displacement in this mode. It is often helpful, although not always essential, to have several actuations at various points. If the plate is actuated with the frequencies corresponding to its eigenvalues, the patterns form that are expected on the basis of the theory. If the plate is actuated with the wrong frequencies, the amplitude of the vibration is very small, since there is no resonance. Also, the amplitude builds up briefly, only to be extinguished again by the actuation. This happens because the actuation frequency and the plates eigenfrequency do not match, with the result that the two rapidly get out of phase.

Table 6.7. Theoretical and calculated frequency ratios for the values given in Table 6.6

Mode Relation	Theoretically	FDM Clammed	FEM Clammed	FEM 758	FEM 3032 $SD = 0$	FEM 3032 $SD = 0.01$
1/2	2.04	2.06	2.04	2.51	2.50	2.50
1/3	3.01	2.87	3.01	4.04	4.01	4.00
1/4	3.66	3.95	3.67	5.07	5.02	5.01
1/5	4.58	4.60	4.60	6.62	6.53	6.52

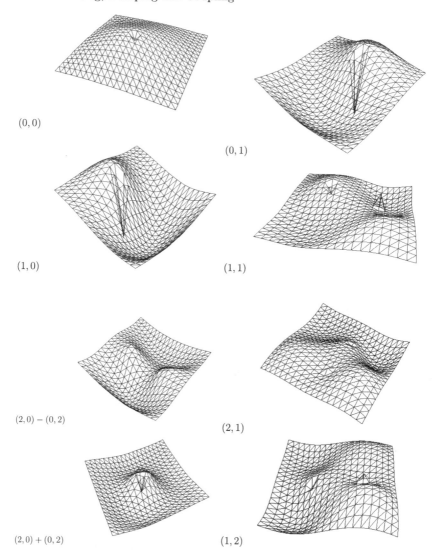

$(0,0)$

$(0,1)$

$(1,0)$

$(1,1)$

$(2,0) - (0,2)$

$(2,1)$

$(2,0) + (0,2)$

$(1,2)$

Fig. 6.19. Diagram of FDM modes. The modal nodes and displacements described above are shown for the frequency values given in Table 6.6: $(0,0)$, $(1,0)$, $(0,1)$, $(1,1)$, $(2,0)(0,2)$, $(2,0) + (0,2)$, $(2,1)$ and $(1,2)$. The nodal lines correspond to the theoretical values. The discontinuities in the plate network are the points of actuation. Here a forced vibration is actuated with the frequency of the mode. With the correct actuation frequency the pattern that is theoretically expected arises in a stable form

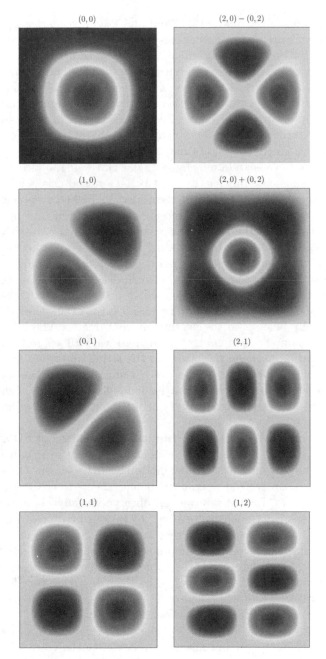

Fig. 6.20. Diagram of modes in FEM. The modal lines and displacements described above are shown for the frequency values given in Table 6.6: $(0,0)$, $(1,0)$, $(0,1)$, $(1,1)$, $(2,0)(0,2)$, $(2,0) + (0,2)$, $(2,1)$ and $(1,2)$. The *nodal lines* correspond to the theoretical values except for $(1, 0)$ and $(0, 1)$. In these cases the *nodal lines* run across the plate and not parallel to the sides

The actuations of the modes $(2, 1)$ and $(1, 2)$, however, are a good example of this. They were created by two actuators providing their stimuli in the central elements $180°$ out of phase but at the resonance frequency. The pattern with the two vibration nodes along either the longitudinal or the transverse direction only occurs with these sites of actuation if the actuation frequency matches this frequency pattern. Otherwise at less suitable frequencies the $(1, 0)$ mode would form, or at higher frequencies the $(3, 1)$ mode. At intermediate frequencies the system would not reach a stable state, but would instead, integrated over a longer period of time, vibrate at the enforced frequency with a small amplitude. We should remember that FDM was designed not to determine modal frequencies, but for transient behavior, and this is where its forte lies. The eigenvalues of plates are determined in FEM by solving a system of eigenvalue equations, which is a good way of tackling the present problem. However, at the same time, some eigenvalues may be missing here that were measured on differentiated sounding bodies. Complex geometries, such as those of percussion instruments, when subject to something similar to a Dirac actuation, can cause reflections between discontinuities in the geometry to create frequencies at the beginning, which are not covered by solving the eigenvalue equation of the system. The piano is an example of this – which also theoretically belongs to the percussion instruments. The brief time during which the hammer is in contact with the string allows reflections of the produced vibration between the bridge of the string and the point of contact with the hammer. These frequencies are a genuine part of the sound of the piano, but have nothing in common with the mode frequencies of the string or of the hammer, and can only be found during about the first two milliseconds (Reuter 1995).

The above diagrams of the plate vibrations in the case of FDM were produced by actuating the plate at the resonant frequency so that it started vibrating. The movement runs "in slow-motion" on the screen. Once the plate is vibrating, it is "frozen" at maximum displacement, to create the situation seen in the diagram. In this way we can observe visually many theoretically well-known phenomena, such as "beating" or various phase relations, as much as we wish. There is a very clear representation, for example, of beating, during which the actuation frequency is set only slightly different from the eigenfrequency of the plate. Since the values are almost identical, the periods during which the motion builds up and dies down again are very long. We can observe the beat in detail, which spans many periods. The pattern of vibration never disappears. Only its amplitude swells and decays.

The discontinuities in the pattern of the vibration come from the points of actuation. These points are not always in phase with the vibration, otherwise no energy could be transferred. In the ideal case the phase difference is $\pi/2$, so that the maximum amount of energy can be transferred.

Table 6.8. Frequencies of various guitar plates. This table is not intended to provide a direct comparison, but instead a relative one (for the mode ratios, see Table 6.9). The modes of Rossing et al. refer to the Western guitar, while those of Richardson refer to the classical guitar. The calculations for FEM were carried out with the aid of the program FEMLAB according to the Mindlin plate theory with the same geometric data that were obtained for the FDM guitar model from a classical guitar

Mode	Richardson	Rossing	FEM
1	41 $(0,0)$	163 $(0,0)$	230 $(0,0)$
2	68 $(0,1)$	276 $(0,1)$	369 $(0,1)$
3	73 $(1,0)$	390 $(1,0)$	495 $(0,2)$
4	114 $(0,2)$	431 $(0,2)$	502 $(1,0)$
5	155 $(1,1)$	643 $(1,1)$	707 $(1,1)$

Table 6.9. Frequency ratios of the first five frequencies from Table 6.8. The values for f_2/f_1 are very similar in all three cases. Then the deviations begin. The FEM calculation reverses the order of the modes $(1,0)$ and $(0,2)$, because the values all lie close together (for the FEM calculation, see Table 6.6)

Modal Relations	Richardson	Rossing	FEM
1/2	1.67	1.69	1.60
1/3	1.78	2.39	2.15
1/4	2.78	2.64	2.18
1/5	3.78	3.94	3.07

Tables 6.8 and 6.9, and the diagrams in Fig. 6.21 for the guitars top plate in its first five fundamental modes of vibration, help to illustrate the modes. Even this simplified top plate (compare Fig. 6.21), without any bracing or bridge, already shows the typical modes of the guitar. Naturally, the high fundamental frequency of 230 Hz is due to the lack of the additional mass of the bracing. The frequency ratio of 1.60 for $\frac{f_2}{f_1}$ is, however, close to the values measured by Rossing et al. of 1.69 and by Richardson of 1.67. The modes $(1,0)$ and $(0,2)$ are swapped over in FEM compared to the reference values, because of the proximity of the eigenvalues of these vibrations. Here we could add many more reference values, but this will be done in the chapter on the guitar. I shall only show now that the argumentation concerning the plate is very simple, and can be transferred to the complicated structures of the guitars. One can expect that in principle they will produce the correct results here the vibration node systems. The calculations were carried out and visualized for the plate part of the MTMS software.

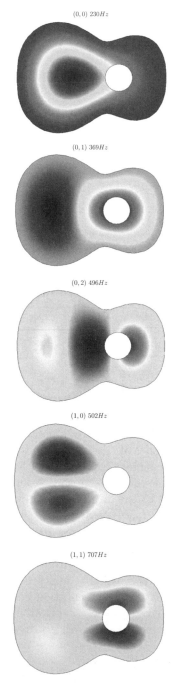

Fig. 6.21. Diagram of the modes of a guitar top plate fixed on all sides (with the sound hole free). The top plate does not posses any bracing or a bridge as yet. The analysis of the eigenvalues is only intended to assist the representation of the system of node lines here. For modes and frequencies, see Tables 6.8 and 6.9

6.6 Damping

6.6.1 Velocity Term as Damping Term
with the Harmonic Oscillator

The damping is generally governed by a velocity term. The overall idea is
to associate the damping with friction, which can only arise when the body
moves. This friction corresponds to the behavior of the material – it is a
plausible reason for the damping. Furthermore, this damping is dependent
upon the frequency. The differential equation of a point vibrator

$$m \frac{d^2u}{dt^2} + D \frac{du}{dt} + K\,u = 0 \tag{6.60}$$

with displacement u, velocity $\frac{du}{dt}$, acceleration $\frac{d^2u}{dt^2}$, mass m, spring constant
K and finally with damping D, has the solution

$$u(x,t) = \hat{A}\,e^{i\omega t} . \tag{6.61}$$

If we put this into the differential equation, we obtain

$$\hat{A}\,e^{i\omega t}\,(m\omega^2 + iD\omega + K) = 0 . \tag{6.62}$$

This produces the eigenfrequency ω as

$$\omega = \frac{-iD \pm \sqrt{-D^2 - 4Km}}{2m} . \tag{6.63}$$

For the non-damped case with $D = 0$, the frequency is

$$\omega_0 = \sqrt{\frac{K}{m}} \tag{6.64}$$

The relationship between the damped case and the non-damped one is therefore

$$\omega = \sqrt{\omega_0^2 - \left(\frac{D}{2m}\right)^2} . \tag{6.65}$$

This means that the frequency of the damped vibration is smaller than that
of the non-damped vibration. In other words, the frequency in the case with
damping, which we always have, depends on the amount of damping. This
can become problematic, if the exact extent of the internal damping we are
considering here is not known precisely.

Here a damping would be advantageous that leaves the eigenvalues of
the plate unchanged. This damping should also depend on the frequency, i.e.
the amplitude of the overtones should die away exponentially along with the
fundamental tone. In the case of the Duffing equation mentioned above, the
eigenvalues do die away with the fundamental tone as determined by

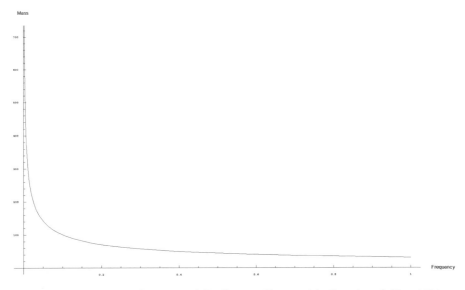

Fig. 6.22. Frequency of a damped Duffing oscillator with $D = 1$ and $K = 1000$, as it varies with the mass $1 > m > 0.001$

$$A(D, m, t) = e^{-(D/2m)\, t} \tag{6.66}$$

If we increase the mass m while keeping the damping constant D the same, so that the eigenvalue is lowered, the damping also changes. If we assume the values $K = 1000$ and $1 > m > 0.001$ for a vibrator, then the eigenfrequencies lie between $1000 > \omega > 31.6$. These values should be regarded as examples. If we now add damping $D = 1$, which we leave constant, the frequency alters if the mass changes as shown in Fig. 6.22.

The change in damping when the mass is altered is shown in Fig. 6.23.

If we combine the two exponential curves, we obtain damping curves which depend on the frequency at various points in time after the start of the vibration, shown in Fig. 6.25.

The curves show an interesting course. The exponential nature of the damping is dependent on the duration of the damping. After a time $t = 10$ ms the damping in relation to the frequency *approaches linearity*. Only afterwards does it become exponential and raises the exponent with increasing duration. Before $t = 10$ m the curvature of the damping curve was actually reversed. The higher frequencies are still more strongly damped than the lower frequencies, but with strong preference for the lower frequencies. But this tendency can be observed for all the curves when low frequencies are used. They are all S-shaped.

The results presented apply to a Duffing oscillator, for which the mass changes, thus influencing the frequencies. Even for a simple system like a damped oscillator, the frequency dependence of the damping behaviour is

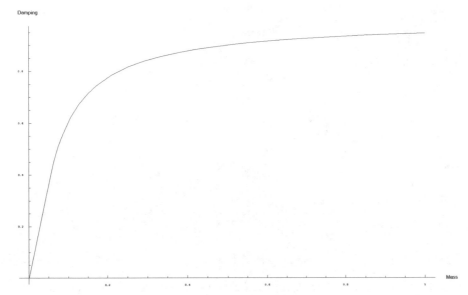

Fig. 6.23. Damping of a Duffing oscillator with $D = 1$ and $K = 1000$, as it varies with the mass $1 > m > 0.001$ after 100 ms. The damping values are shown dependent on the maximum amplitude 1

transient and converges slowly to a value (in Fig. 6.24 in direction to larger time points). In opposite to the simple oscillator, a guitar is made of three dimensional plates, which consist mainly of bending waves. These are described by differential equations of fourth order (see above). The Duffing equation is, however, a second order differential equation, just like the longitudinal waves in the plate. So there are analogies here. The damping of a harmonic oscillator very its vibration frequencies. This behaivour should be expected from a vibrating plate, too. But it is our aim, to have a damping term for plate vibrations, with leaves the vibrational frequencies unaltered. This allows several damping values for modulation of realistic sounds. So the damping of wooden plates can be determined in a way of adjusting the damping values of the model to those of the real plate.

6.6.2 Force Term as Damping Term with Bending Waves

In this book a damping for bending waves is proposed that does not change the eigenvalues of the system. With a damping constant R, the following applies:

$$\frac{\partial \triangle u_z}{\partial t} = -\frac{E * K^2}{\rho} * R \frac{\partial \left(\dfrac{\partial^4 d_z}{\partial \left(x + \frac{\partial^2 d_x}{\partial x^2} \right)^4} + \dfrac{\partial^4 d_z}{\partial \left(y + \frac{\partial^2 d_y}{\partial y^2} \right)^4} \right)}{\partial t} \tag{6.67}$$

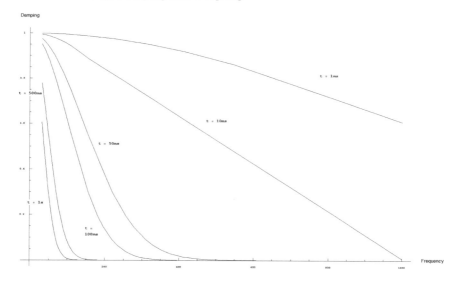

Fig. 6.24. Damping of a Duffing oscillator with $D = 1$, $K = 1000$ and $1 > m > 0.001$ as it varies with the frequency when the mass changes. The damping values are shown dependent on the maximum amplitude 1. The curves represent the course at the points in time $t = 1\,\text{ms}$, $t = 10\,\text{ms}$, $t = 50\,\text{ms}$, $t = 100\,\text{ms}$, $t = 500\,\text{ms}$ and $t = 1\,\text{s}$

Here the velocity of the bending wave $\frac{du^z}{dt}$ is changed by $\frac{d\triangle u^z}{dt}$. The basis for this is the derivative of the force $\left(\dfrac{\partial^4 d_z}{\partial(x+\frac{\partial^2 d_x}{\partial x^2})^4} + \dfrac{\partial^4 d_z}{\partial(y+\frac{\partial^2 d_y}{\partial y^2})^4}\right)$ with respect to time. If we choose $k = -\frac{E*K^2}{\rho}$, and call the change in velocity $\triangle v$, we can write the above equation like this:

$$\triangle v^z = k \, R \, \frac{\partial F}{\partial t} \,, \tag{6.68}$$

with F being the force.

The idea behind this is derived from the bending motion. The internal friction, which we are concerned with, is regarded here as a type of "groaning" on the part of the plate. If a force impinges on an element, this means that at that point in the element there is a bending, i.e. the plate is curved at this location. If this curvature changes, it means that the wood (or metal) has to do some work. The internal shape changes. The change of shape is viewed here as the cause of a loss of energy. This loss is independent of the velocity. A change in shape can occur without the element moving, for example when the neighboring elements alter their position. Then the curvature of the plate changes, too, at this place, and with it the force. So this temporal change in force is seen as the cause of a loss in energy. But of course this loss is expressed in the velocity of the element. So if the velocity of the element is zero, there is no loss of energy either. This ensures that only those elements which have

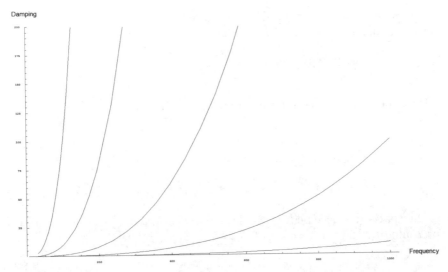

Fig. 6.25. Frequency-dependent damping with $t = 1$ for various values of R. For wood $R = 10^{-4}$, and for metal $R = 10^{-7}$. The intermediate forms between wood and metal are also audible as such

a velocity and a curvature will dissipate energy. This is because the change in curvature at an element at rest can only be seen as a shift of the neighboring elements. These elements, however, therefore also experience a change in their curvature. So every change in curvature incurs a loss of energy, but changes in velocity only do so if there is a concomitant change in curvature.

This damping has two advantages. One the one hand, the eigenvalues of the plate are independent of the damping, and they do not become smaller if the damping increases, as is the case with the damping of the velocity of a Duffing oscillator. On the other hand, here we have the strict superposition of all vibrations that occur in the body. There is no coupling of the modes because of this damping. This can happen when there is velocity damping. If, for example, two vibrations are superimposed, one of a low frequency and one of a high frequency, then every element in the plate that has a large velocity owing to the high-frequency vibration is severely damped. But this damping also affects the fundamental frequency. Since the eigenvalues depend on the damping, the fundamental frequency is somewhat lowered, just because another type of vibration is occurring next to it in the plate. This is not the case with damping of the force, as the eigenvalues of the vibrations do not depend on the damping.

The extent of the force damping $A(\omega)$ in dependence of the angular frequency ω is given by

$$\mathcal{A}(\omega) = \tau \, R \, \omega^3 \tag{6.69}$$

with R being the strength of damping and the damping dependent material constant τ.

The independency can be derived from the sine-wave vibration. In this case the bending displacement for a plate with free edges is given by (Wagner 1947)

$$u^z(x,t) = e^{i\,\omega\,t}\left(\alpha\,e^{k\,x} + \beta\,e^{-k\,x} + \gamma\,e^{i\,k\,x} + \delta\,e^{-i\,k\,x}\right) \qquad (6.70)$$

Here the dependence on time is represented by

$$u^z(t) = e^{i\,\omega t} \qquad (6.71)$$

and the spatial distribution by

$$u^z(x) = \left(\alpha\,e^{k\,x} + \beta\,e^{-k\,x} + \gamma\,e^{i\,k\,x} + \delta\,e^{-i\,k\,x}\right) \qquad (6.72)$$

The spatial displacement of the bracing corresponding to this equation is represented by the two terms

$$\gamma\,e^{i\,k\,x} + \delta\,e^{-i\,k\,x} \qquad (6.73)$$

which correspond to d'Alamberts solution for a string, with one wave from right to left and one from left to right. There are also the terms

$$\alpha\,e^{k\,x} + \beta\,e^{-k\,x}\,, \qquad (6.74)$$

which represent a type of end correction, as a bending which moves sine-like in time, is spacially strained in a way of being teared towards the ends of the geometry.

From the above equation is is obvious, that the spacial maximum displacement curve is quite complicated. It follows from the discussion of just one frequency ω of a sine wave,

$$A = \hat{A}\sin(\omega x)\,, \qquad (6.75)$$

which is deformed in space. If we let aside this deformation, then the curvature can be described as

$$\frac{\partial^2 A}{\partial x^2} = -\omega^2 \hat{A}\sin(\omega x) \qquad (6.76)$$

The curvature is a derivation of second order with respect to space. We can neglect the end correction of the bending movement, as it is the spacial curvature changes at each place, and so is spread over a larger spacial distance. In both cases, with and without end correction, a wave length π or $\pi/2$ and multiples of this are present. In the case of the end correction over the whole geometry of the rod the same curvature is given like in the case of no end correction.

This curvature is differentiated a second time according to the method proposed here, now with respect to time, to calculate the change of curvature. The next derivative, which has the same angular frequency ω in the sine term

$$\frac{\partial A}{\partial t} = -\omega^3 \hat{A} \sin(\omega t) \tag{6.77}$$

gets us the ω^3-dependence of frequency and damping.

To be able calculate sounds using these algorithms, we still have to determine the sound decay of the plate, its maximum time of sounding. A simulated sound of knocking on wood continues to sound for about 50–100 ms. With metal the sound can last for several seconds. To simulate the whole decay time, another pure velocity term is now added. It calculates the current velocity as a percentage.

$$v_{\text{damped}} = v_{\text{undamped}} * k \tag{6.78}$$

For wood, a value of $k = 0.9995$ is used, and for metal $k = 0.999995$.

Figure 6.26 shows five calculated plate sounds. Each of the sounds lasted for five seconds and was calculated with a 96 kHz sampling rate for a 40 cm by 40 cm plate with a thickness of 3 mm, which is fixed on all sides. It was struck in the center, which means that an element was displaced as the initial condition and the calculation was then run. The calculations were performed for the plate part of the MTMS software.

The sounds were produced by integrating over the whole surface at every point in time. The velocity of the plate parts and the plate node points are integrated with respect to a point in the room, which has a distance of 1 m perpendicular to the guitar top plate. Interestingly, it was essential to use not only a spatial distance of travelling waves of the individual plate elements. Obtaining realistic sounds also required consideration of the fact that the element with the shortest path contributes the greatest proportion of energy to the sound. The greater the distance of an element from the virtual microphone, the more its sound had to be damped. It holds

$$P(\mathbf{e}) = \begin{cases} 1/(|\,e_x - l_x/2\,| + |\,e_y - l_y/2\,|), & -1 > e_x > 1 \quad \& -1 > e_y > 1\,; \\ 1, & -1 < e_x < 1 \quad \& -1 < e_y < 1\,; \end{cases} \tag{6.79}$$

if P is the proportion of the elements $\mathbf{e}$ in the radiated sound and $\mathbf{l}$ denotes the dimensions of the plate. Otherwise, with a pure sine wave actuation, no sound would be produced at all. While namely one part of the surface vibration provides a positive component to the sound, the other part of the plate contributes an equal negative component. While at a distance of 1 m, the difference in the spatial distance of travelling waves was not audible, a sound without any energy reduction through spatial distance would have nothing in common any more with a sound from a real plate.

Figure 6.27 shows the spectra of the two sounds where $R_1 = 10^{-7}$ and $R_2 = 10^{-6}$, both with $k = 0.999995$ in both cases, integrated over the whole duration of 5 seconds. Only here does it begin to become clear that the overtones are very much reduced in the case of R_2 compared to R_1. What is interesting is that the amplitude of the fundamental tone remains exactly the same. Both

a)

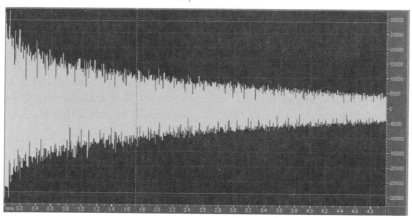

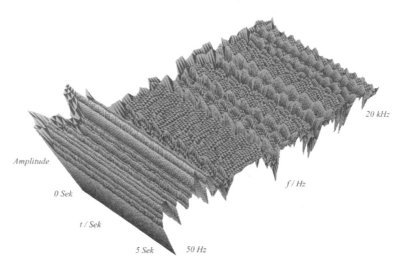

Fig. 6.26. Time series and their wavelet transformations of the calculated plate sounds for a plate of 40 cm by 40 cm, 3 mm thick and with various damping values of R and k. (a) $R = 10^{-7}$, $k = 0.999995$ (b) $R = 10^{-6}$, $k = 0.999995$ (c) $R = 10^{-5}$, $k = 0.99995$ (d) $R = 10^{-5}$, $k = 0.9995$ (e) $R = 10^{-4}$, $k = 0.9995$. Wavelet transformations of 50 Hz–20 kHz over 5 seconds. The calculation and visualization were performed using the MAS software (see chapter on software)

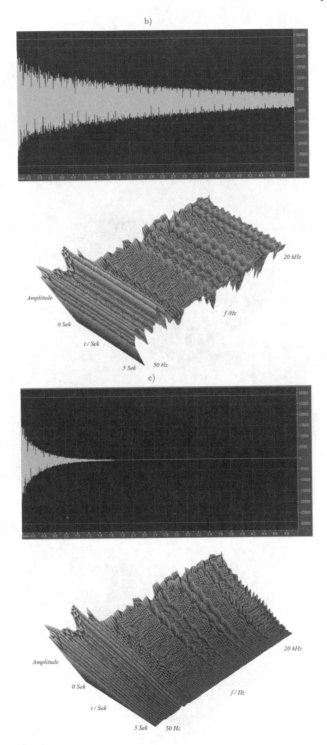

Fig. 6.26 (*continued*)

126   6 Bending, Damping and Coupling

d)

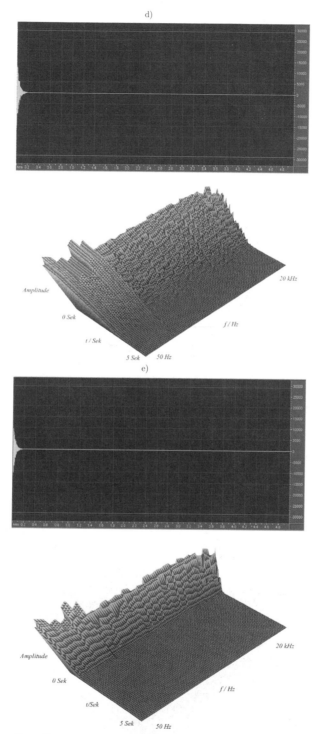

e)

Fig. 6.26 (continued)

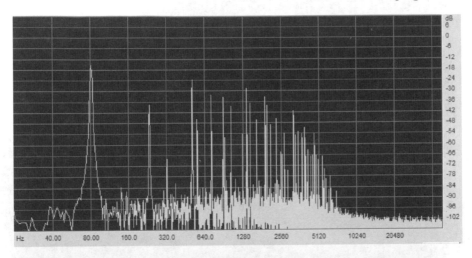

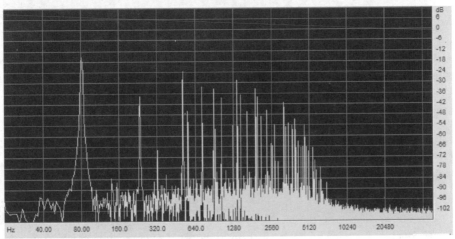

Fig. 6.27. FFT spectra of the two plate sounds with $R = 10^{-7}$ (*top*) and $R = 10^{-6}$ (*bottom*), with $k = 0.999995$ in both cases, amplitude-normalized and integrated over the 5 seconds. Only here does the stronger overtone damping become apparent, which is difficult to identify in the time-dependent wavelet images

sounds were normalized, i.e. the maximum time-series amplitude occurring was set to $\pm 10^{15}$ or this 16-bit resolution. This largest value naturally applies at the beginning, before any damping has started. The difference between R_1 and R_2 has practically no effect on the amplitude of such a low frequency as 82 Hz. This means that the damping values are very comparable and deliver the desired results.

Table 6.10 contains the first five strong eigenvalues of this plate for the various damping values.

Table 6.10. Frequency values for the five lowest of the strongest eigenfrequencies of a plate 40 cm by 40 cm with a thickness of 3 mm, for various damping values R and k. The force damping does not affect the eigenvalues, even though R was changed by a factor of 1000. The increase to $k = 0.9999995$ from $k = 0.9995$ only corresponds to a factor of 1.0005. If k is increased more the eigenvalues change considerably

$R = 10^{-4}$ $k = 0.9995$	$R = 10^{-5}$ $k = 0.9995$	$R = 10^{-5}$ $k = 0.99995$	$R = 10^{-6}$ $k = 0.999995$	$R = 10^{-7}$ $k = 0.999995$
82.031 Hz	82.031 Hz	82.031 Hz	82.031 Hz	82.031 Hz
235.843 Hz	235.843 Hz	235.843 Hz	235.843 Hz	235.843 Hz
323.731 Hz	323.731 Hz	323.731 Hz	323.731 Hz	323.731 Hz
511.229 Hz	511.229 Hz	511.229 Hz	511.229 Hz	511.229 Hz
558.106 Hz	558.106 Hz	558.106 Hz	558.106 Hz	558.106 Hz

6.7 Coupling

The couplings between the various parts of the guitar were treated like a mass-spring system. In the literature on contact mechanics (Wriggers 2002) a distinction is made between two possible formulations for contact. On the one hand, the method of the Lagrange multiplier is used, while on the other the penalty method is applied.

In classical contact problems the first thing is to recognize the contact case. A ball falling onto a plate (Bensoam 2003), is in contact with the plate for a certain period of time, and is then repelled and loses its contact with the plate again. So a contact condition can be formulated.

$$F_{\text{external}} \begin{cases} f(\mathbf{u}), & \text{with contact } A \leq 0 \, ; \\ 0, & \text{without contact } A > 0 \, . \end{cases} \tag{6.80}$$

If A is the distance between the bodies, contact occurs when this distance is less than or equal to zero. Then the force acting on the body $F_{\text{external}} = f(\mathbf{u})$, i.e. a function of the displacements $\mathbf{u}$ of the bodies.

Both possible types of contact, the Lagrange multiplier and the penalty method, take the energy as their starting point. In a mass-spring system the energy $E = K + T$ where K is the kinetic energy and T the potential energy,

$$E = \frac{1}{2}m \left(\frac{d\mathbf{u}}{dt}\right)^2 + m \, k \, \mathbf{u} \, , \tag{6.81}$$

i.e. dependent on the displacement $\mathbf{u}$.

With the Lagrange multiplier, the external energy term is added as the transmission of force:

$$E = \frac{1}{2}m \left(\frac{d\mathbf{u}}{dt}\right)^2 + m \, k \, \mathbf{u} + \lambda \, A(\mathbf{u}) \, . \tag{6.82}$$

Since $A(\mathbf{u})$ is the distance between the bodies, that is a displacement, λ must be a force. The calculation of λ now depends on the type of coupling.

In FEM this equation must be varied (usually at this point one assumes an acceleration due to gravity, so that the energy becomes $E = \frac{1}{2}k\,\mathbf{u}^2 - m\,g\,\mathbf{u} + \lambda\,A(\mathbf{u})$ with $g = 9.81\,\mathrm{m/s^2}$ for gravity. Varying this equation leads to a value of $\lambda = kA - m\,g$ and corresponds to the force exerted externally.

The second possibility is the notion that when the body makes contact it is coupled as if via a spring with a spring constant θ. The energy observation then becomes

$$E = \frac{1}{2}m\,\left(\frac{d\mathbf{u}}{dt}\right)^2 + m\,k\,\mathbf{u} + \frac{1}{2}\theta\,A(\mathbf{u})^2 . \qquad (6.83)$$

Another way of realizing this, albeit a difficult one, would be to treat the two bodies from the point in time of the contact as if they were only one body. For deformation contacts of round or oval bodies, this would need a great deal of work. This is because in such cases the contact surface is changing continuously, with the result that a new body has to be assumed for every time-step. For contacts where the contact surface neither enlarges nor shrinks, this becomes of interest when a comparison is to be made between this contact and a transmission of a force. In the former case the contact surface is always equally displaced and the surfaces are treated as if they were located within the body. In the latter case, however, no "fusion" is supposed, and the force transmission can take on nonlinear forms.

With FDM an examination of the energy is of no immediate interest. The contact is calculated as a force. In the simplest case a spring constant is assumed, which couples the parts. We then have

$$F^{\mathbf{e}} = F_{\text{surrounding}} + F_{\text{external}} = F_{\text{surrounding}} + \theta\,\mathbf{u}^{\mathbf{e}_{\text{contact}}} , \qquad (6.84)$$

i.e. the force $f_{\text{surrounding}}$, against which the neighboring elements in the body push on the node point $\mathbf{e}$ plus the force arising from the displacement of the contact point $\mathbf{e}_{\text{contact}}$ with the result that the coupling is viewed as being damped by an amount θ.

This damping is an assumption that is difficult to check experimentally. For this reason the following procedure was followed. Once the geometry was complete, the damping values for the preliminary investigations of the plate were entered. If the value assumed for θ were too large, the calculation would overflow after some time after the string was plucked. This happens for the simple reason that the neighboring elements receive a certain amount of transferred energy. The total energy of the system increases continuously if θ has been given an unrealistic value, so the procedure overflows. Below the critical value of θ the coupling works perfectly smooth. This does not mean, however, that the precise value is known. Now, a reduction of θ in the range below the critical value only means that the individual parts of the guitar (such as the ribs or the back plate) pick up more energy, and pass it on to the other parts. This does not affect the actual type of vibration. The radiation of these types of vibrations away from the guitar produces the overall sound. So θ only affects the mixing of the audible portions of the sound from the various parts of the guitar, not their characteristic behavior.

Furthermore, the coupling was carried out using the angular method described above. In the case of coupling between the back plate and the ribs, and between the ribs and the top plate (and vice versa), the angle is 90 degrees. So here a complete conversion takes place from transverse waves into longitudinal waves, and vice versa. Again, this is easy to imagine. A transverse wave running along the top plate and reaching the edge pulls this edge either upwards or downwards, in the direction of the top plate normal. For the ribs, however, the direction of this top plate normal is their longitudinal direction. Since the ribs are regarded as a curved plate whose x direction runs along the ribs, and around the top plate and the back plate, a z-displacement of the top plate becomes a y-displacement in the ribs. The energy of the bending wave is thus turned into the energy of a longitudinal wave. If this rib displacement now reaches the back plate, the y-displacement of the ribs in turn affects the normal of the back plate. The latter is similar to the top plate in its z-direction, along which the bending waves run. So here the wave is converted back into a bending wave. The energy of the top plate is transferred to the back plate via the ribs in a "subterranean" fashion. We can call it "subterranean" because the longitudinal waves in the ribs cannot add any energy to the sound being radiated away. And apart from this, the longitudinal wave is considerably faster than the bending wave, so the first displacement of the back takes place via the ribs, before the coupling via the air inside the guitar can affect the back plate. A detailed representation of the initial transients of the guitar will be found in the section on results later in this monograph.

The coupling of the top plate to the ribs is in its turn brought about by a transformation. A longitudinal wave in the ribs in the y-direction, which gets as far as, say, the back plate, finds a curved surface there. The curvatures indicate one portion of the energy in the back plate in its x- and y-directions. If the wave arrives at the greatest bulge or the waist, all its energy is transferred to the y-bending wave of the back plate. If the wave reaches the lower or upper side of the guitar body, it is completely transferred to the x-direction of the back plate. In between, it is distributed proportionately to the x- and y-directions. This distribution again takes place via the curvature that was calculated above.

7

Results

The virtual guitar that is constructed here allows a huge number of possible investigations. It is not only possible to try out all existing types of wood. One can also investigate materials that do not yet exist, for example one produced using a special carbon composite. Furthermore, one can make changes to the geometry of the guitar to study their effects on the sound at any time, e.g. reducing the height of the ribs, introducing a cutaway (a recession of the body improving the mobility of the left hand) at various locations, shifting the bridge in different directions, altering the length of the strings or the head plate, using any type of bracing or changing the thickness of the top plate, back plate, ribs or neck, various shapes for the sound hole, etc.

In addition, it is possible to effect "changes" that allow us to draw conclusions about the real behavior of the guitar. For example, the effect of longitudinal vibrations of the strings can be switched on and off. This enables us to determine the order of magnitude of these parameter changes. The bridge can be enabled to vibrate as a coupler in three dimensions, or six, or only one, and the sound hole can be open or closed. And all the couplings can be switched on and off, to allow investigation of coupling in one or both directions. For example, coupling from the top plate to the ribs can be allowed, while coupling in the opposite direction is not. In this manner it is possible to test the extent to which the individual parts of the guitar are dependent on one another. So we can look into the question of whether the feedback between the ribs and the top plate really leads to significantly different patterns of vibration in the top plate, or whether this feedback is only of lesser importance.

In addition, damping can be added at any location. How much is the sound influenced by the fact that part of the back of the instrument rests against the guitarists body, or that the ribs are damped by the players legs? What influence is exerted by the left hand pressing on the fingerboard, for example when bar chords are played, where the hand presses strongly on the neck?

Finally, one can test which plucking technique is to be created by each string geometry. The apoyando and tirandu techniques sound different, which on the one hand is a result of the displacement of the string before it is

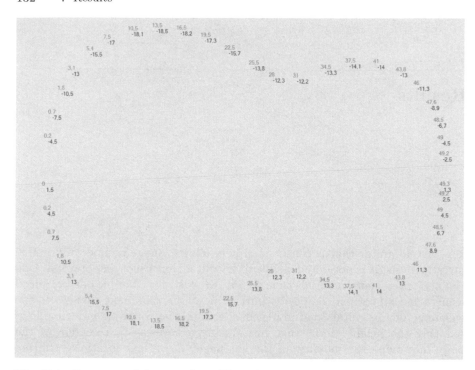

Fig. 7.1. Geometry of the top plate. The values are the x and y coordinates in the guitars coordinate system with the normal vector n

plucked, while on the other it depends on the movement of the finger as it plucks the string. In this case the noise from the finger sliding along the string before it is plucked is also important. This sound is very characteristic of the guitar, because it cannot occur in the harpsichord or the piano. In addition, it contains not only noise, but also part of the later overtone spectrum of the tone, or the eigenvalues of the guitar body. If the finger noise is left out an important part of the tone is missing. Modelling this noise can provide further information about the type of plucking. If the desired noise is generated, one can be pretty sure of having identified the course of the plucking correctly.

We cannot go into all of these many possibilities in depth here. I shall therefore pick out some interesting individual aspects. The most important of these is the transient.

7.1 Parameters of the Guitar Model

The guitar was modelled using parameters typical of the instrument, which will be given here and which are valid for all the other calculations, unless otherwise stated.

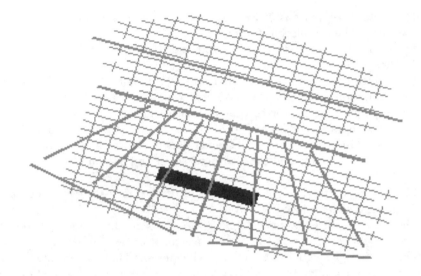

Fig. 7.2. Bracing, bridge and sound hole of the top plate

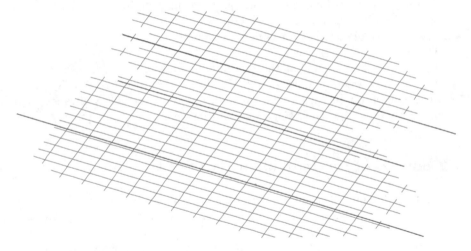

Fig. 7.3. The bracing of the back plate consists of three bars

- The elasticity modulus of the plates of the guitar is divided into Youngs modulus along and across the grain of the wood. The values $E_x = 13.79^9$ and $E_y = 10.77^8$ were used throughout, i.e. a ratio of 1 : 12.8, which is the average of the values given in the literature, of 1 : 8.01 (Jahnel 1986), 1 : 12.8 (Schneider 1997a) and 1 : 15.45 (Fletcher and Rossing 2000). The density of the woods $\varrho = 1175\,\mathrm{kg/m^3}$. Their damping values are assumed to be those for wood, i.e. 10^{-4} or damping and 0.9995 for damping based on velocity. The Poisson ratio $\mu = 0.3$.

- The guitars top plate is 49.3 cm long and symmetrical about the guitars normal n. The top plate is 3 mm thick.
- The sound hole has a radius of 4.5 cm and its center is given in the coordinate system as $S_m = \{33.5\,\text{cm}, 0\,\text{cm}, 0\,\text{cm}\}$.
- The bridge has a length of 18.5 cm and a width of 3.2 cm, and in a first approximation its thickness is assumed to be a constant 0.4 cm.
- The top plate has a Torres bracing consisting of a seven bar fan, two transverse bars at the lower end of the belly and two thicker supporting bars around the sound hole. The seven bars of the fan and the bars in the belly are 5 mm thick, and those around the sound hole are 2 cm thick (see Fig. 7.2).
- The bracing on the back plate consists of three transverse bars. They have a thickness of 2 cm (see Fig. 7.3).
- The swallow-tail is assumed to be a block 2.5 cm thick, and the wooden block at the lower end of the body is assumed to be 2 cm thick.
- The ribs have a height of 9.8 cm and a thickness of 3 mm. The values for the curvature have already been given above in the section on the curvature calculation. Figure 7.1 gives the ribs coordinates.
- The neck is 50 cm long, 6 cm wide and 2.4 cm deep at its thickest point, and is rounded off towards the outside.
- The velocity of sound in air is taken to be $c = 330$ m/s.
- The guitar has a string length von 65 cm.
- The strings are fastened at the relevant places at the bridge and the nut. The angle between them and the bridge when they are at rest is 45°. Their fundamental pitches arise from the assumption of suitable values for their linear densities and their fundamental tensions.

7.2 Sounds of the Individual Parts of the Guitar

The exact behavior of the parts of the guitar within the initial guitar transient is not yet known. I shall tackle a noisy, percussive component of the sound, which precedes the harmonic sound (Richardson 2003).

In previous studies (Bader 2002a) the initial transients of the most important types of instruments have been described, including the guitar. It was shown that the apparently chaotic commencement of the sound can often be traced back to just a few natural laws. In the case of the guitar it was found that these were generally the vibrating eigenvalues of the guitar body and strong fluctuations of amplitude in the individual partials. This characteristic of a tone is independent of its volume. For any given pitch on a particular string on a particular instrument, the guitarist can only vary the volume. He has no influence over the initial transient lasting approx. 50 ms, which is different than playing the violin or the saxophone, where the tone can still be influenced after it has started. The initial transient of a guitar tone always develops in exactly the same way, independent of its volume. This astounding constancy in the events of the initial transient of a guitar makes it very suitable for investigation, since we can be sure that the transient found is not dependent upon the displacement of the string assumed at the beginning of the modelling calculations.

7.3 Couplings

Figure 7.4 on the previous page shows the temporal development of the initial transient of a tone at the pitch e^1, generated by plucking the open high E string. The form adopted by the string is a triangle, and the point of plucking is one fifth of the way along the length of the string as seen from the bridge shown in Fig. 7.5.

These time series already reflect the basic pattern of the functions of the individual parts of the guitar. The top plate starts vibrating immediately. It more or less reflects the development of the displacement of the string at the bridge. Since the algorithm covers the coupling between string and bridge in two ways, which will be described in detail in the next section, the reader is referred to that chapter.

Once the top plate has started vibrating, it couples directly to the air space, and slightly later to the ribs. The coupling to the ribs can only follow when the wave in the top plate reaches the ribs. There the initially purely transverse vibration is converted into a longitudinal one, since the coupling between the top plate and the ribs has an angle of 90°. However, the ribs do not only vibrate longitudinally, since their curved structure continually changes the longitudinal waves into transverse ones. So a very quiet "glitter" arises initially at the ribs. But this vibration is radiated away, since it occurs in the z direction of the ribs, that is, it is a bending wave.

Figures 7.4 and 7.6 represent the relevant time series for the first few milliseconds of the initial transient. We should remind ourselves that the amplitudes of the parts of the guitar have all been standardized in these diagrams, to allow a visual comparison.

7.4 Spectra of the Parts of the Guitar

The diagrams of the spectra allow the eigenvalues to be determined, which are added to the sound during the initial transient and can be ascribed to the various parts of the guitar. Tables 7.1 and 7.2 show the eigenvalues of the individual guitar parts, the body resonances found in the initial transient of the tone e^1, and the spectrum and frequencies of the whole guitar. The spectra were evaluated up to 6000 Hz. The frequencies of the individual parts of the guitar (see Fig. 7.8) and of the whole body (see Fig. 7.7) were obtained in a transient analysis by displacing one element of the relevant part at the beginning. Thereupon, the parts of the body start to vibrate. The radiation takes place as described above, integrated over the surface. The time series thus

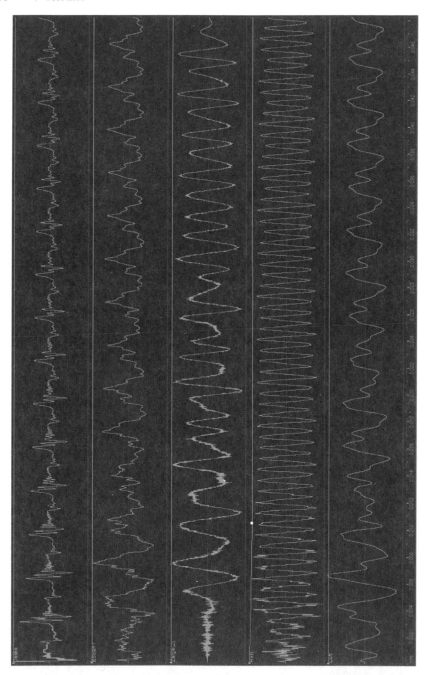

Fig. 7.4. Time series in the first 50 ms of the guitar tone, calculated using FDM. The parts of the guitar are shown individually. From *top* to *bottom*: top plate, back plate, ribs, neck, enclosed air. The amplitude values have been standardized at 100

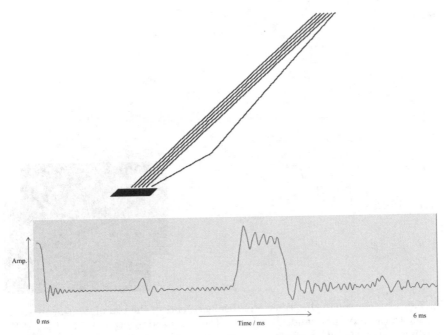

Fig. 7.5. *Above*: Displacement of the string at the starting point of the calculation. *Below*: Time series of the displacements of the string at the bridge point during the fist 6 ms

produced were subjected to a windowed FFT. As the plucking is practically a point actuation, all the eigenvalues are stimulated, up to the case in which precisely this value at the point at which the string in plucked has a vibration node. But since the elements are finite, this can never be quite true. So only all the possible eigenvalues are actuated. Their relative amplitude strengths can be assumed to be approximately correct. Strong frequencies also appear clearly with this method. However, the "strength" of a mode is also dependent on its damping during the transient. The contribution of the individual modes later on in the tone also still depends to a great extent on the location of the plucking. This is a decisive factor in the case of the top plate, which is stimulated by the string at the bridge. Generating a spectrum of the guitar body with exact data for the amplitudes therefore always depends on the actuation. And precisely for such a complicated system as the guitar and with the transient found here, it will not necessarily be possible to draw conclusions from a precise determination of the amplitudes of the body eigenvalues caused by external actuation, e.g. by a sine-wave vibrator. One tendency will become apparent, but for precise analysis the transient of the whole body must be taken into account.

The tables tell us, that the frequencies of the single guitar parts do not correspond to the eigenvalues of the closed guitar body. Partly a rough assignment

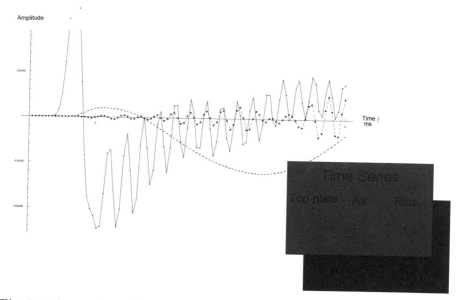

Fig. 7.6. Comparison of the transverse displacements of the top plate, the air space adjacent to the top plate, and the ribs for 1 ms, integrated over $\mathbf{D}_z$, $\mathbf{L}_z(0)$ and $\mathbf{Z}_z$ according to the algorithm given. The top plate reacts very rapidly, and then the air coupling begins, followed by the coupling of the ribs

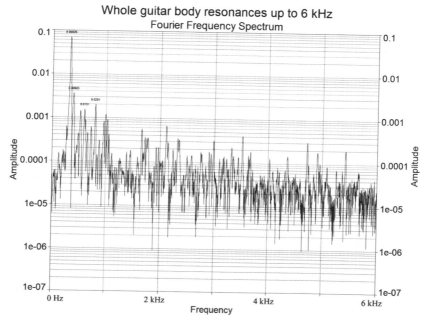

Fig. 7.7. Spectrum of the guitar body up to 6000 Hz

is still possible, most of the time, a reasonable comparison is no longer possible. This is known for the fist three modes of the guitar which are called combinational modes. These modes come into place by the interaction of top plate, inclosed air and back plate. But even the coupling of these parts is not enough to obtain the measured values. So, it is essential always to consider the whole guitar. We should remind ourselves again that the modes discussed here are only one aspect of the sound. What is important is the transient development of the sound and the involvement of the individual parts of the body in it. The eigenvalues of the body, which generate the percussive sound in the initial transient, alter this "knocking" at the beginning so little that here their precise frequencies contribute little to the character. The exact development of the initial sound, however, and the question of the interaction between the string mode and the bodys eigenvalues and their temporal development, are, by contrast, decisive factors affecting the character of the sound. The contributions of the various parts of the body to these developments provide results that allow us to predict the changes to the sound from the body. This is the actual objective of this monograph, rather than finding the exact frequencies of the modes.

Table 7.1 gives the values of the frequencies found in the initial transient of the tone e^1 on the guitar in the sounds radiated by the various parts of the guitar and which do not belong to the overtone spectrum of the string. The values in bold type are the frequencies found in several parts; they are separated from the others by the subdivisions.

The top plate and the back plate exhibit far fewer frequencies in the range up to 6000 Hz than do the ribs or the enclosed air space inside the guitar. The lowest frequencies that occur are naturally close to the fundamental tone, which in this case is 330 Hz. They occur at 399.90 Hz to 401.37 Hz, and from 450.538 Hz to 454.1 Hz, and can be ascribed to the combination modes. But the mode around 400 Hz, for example, does not occur in the top plate at all. And the values around 1020 Hz and 1100 Hz are missing in the back plate. Here the roles played by the individual parts of the guitar are already becoming evident, which become visible in the development of the transient. The top plate is the first part to react. This means that the body eigenvalues of the top plate are much less pronounced. Their sound is much brighter and more direct. It is not responsible for subtleties in the tones of the guitar.

Furthermore, we find four frequencies between approx. 3917 Hz and 4890 Hz in the ribs, enclosed air space and neck. These frequencies lie adjacent to those of the tone e^1 in the diagrams of the spectra of these parts of the guitar, that is, next to the overtones at the right-hand boundary of the spectra. As can be seen from the tables above, these are eigenvalues of the whole guitar body, which exist alongside the values of the overtones of the e1 tone played on the instrument. For this reason they are also stimulated, and contribute to slow beats in the tone. Another common value of 2093.57 Hz to 2096.64 Hz occurs between the top plate and the neck. This tone is a common mode of these parts.

Table 7.1. Frequencies of the modes occurring during the initial transient, which are not eigenvalues of the string. The frequencies given only apply to their measuring points, and not to the cause of these frequencies. It will become clear that often a precise allocation to individual parts of the guitar is no longer possible. The properties of the instruments parts (top plate, ribs etc.) are not adopted in the vibrating guitar body as a whole. It vibrates at eigenvalues other than those of the isolated parts

Top Plate	Back Plate	Ribs	Inclosed Air	Neck
	400.48	**401.37**	**399.90**	
452.64	**450.538**	**454.1**	**452.64**	**453.33**
		583.01		
			739.75	
1021.73		**1020.96**	**1022.5**	**1023.36**
1106.88		**1107.46**	**1099.97**	
1252.13	**1271.52**			
1437.4				
1592.74		1606.94		
	3133.73	1935.07		1443.36
	4440.29			
2093.57				**2096.64**
2198.78	4750.66	2586.91		
2414.11	4900.9	2965.34		
	6057.22	3577.15		
		3919.97	**3917.76**	**3916.7**
			4246.56	**4243.3**
		4573.25	**4573.25**	**4569.98**
			4899.94	**4896.67**
		5526.72	5493.12	5223.36
			5808.1	
			6123.07	
			6435.07	

Let us now look at Table 7.2, in which the eigenvalues of the individual parts of the guitar are shown. In the lower frequency range all the values are shown. In the upper range I have only given those values that occur in the frequency ranges of Table 7.1. So we are looking for the causes of the body modes in the initial transient and how they can be allocated to the individual parts of the guitar. It is easy to see that such an allocation is impossible. Sometimes one frequency appears to be close to one of the frequencies listed in Table 7.1. But it is not possible to allocate them even approximately.

Instead, let us examine Table 7.3, and here we can recognize the links immediately. This table does not show the modes of the individual guitar parts that have been investigated separately, but instead the modes of the whole body corresponding to the modes found in the initial transient. The diagram below represents the whole spectrum for the guitar body. It clearly shows

Table 7.2. Eigenfrequencies of the individual parts of the guitar. Frequencies of up to approx. 6 kHz were taken into account

Top Plate	Back Plate	Ribs	Inclosed Air	Neck
393.6	230.4	297.6	114.0	652.8
470.4	422.4	499.2	182.4	710.4
633.6	480.0	537.6	268.8	883.2
720.0	662.4	643.2	355.2	1190.4
873.6	912.0	700.8	403.2	1612.8
960.0	988.8	768.0	460.8	1900.8
1056.0	1123.2	835.2	585.6	1996.8
1084.8	1209.6	902.4	710.4	2131.2
1152.0	1296.0	960.0	739.2	2169.6
1238.4	1363.2	1075.2	844.8	2457.6
1440.0	1488.0	1123.2	969.6	2726.4
1488.0	1545.6	1180.8	1084.8	2860.8
1651.2	1612.8	1209.6	1152.0	3369.6
1689.6	1680.0	1248.0	1190.4	3417.6
1737.6	1766.4	1296.0	1315.2	3475.2
1795.2	1804.8	1344.0	1382.4	3580.8
1891.2	1891.2	1401.6	1430.4	3772.8
2025.6	2025.6	1488.0	1536.0	3964.8
2054.4	2140.8	1574.4	1632.0	4060.8
2092.8	2179.2	1651.2	1718.4	4320.0
2121.6	2313.6	1718.4	1776.0	4368.0
2265.6	...	1766.4	1833.4	4473.6
2313.6	...	1795.2	1881.6	4588.8
2361.6	...	1872.0	1900.8	4627.2
2400.0	2400.0	1891.2	2112.2	4771.2
2448.0	2448.0	1958.4	2208.0	4809.6
2534.4	...	1958.4	2275.2	4982.4
	3024.0	2016.0	2342.4	5126.4
	3139.2	...	2457.6	5308.8
	3273.6	2390.4	2515.2	5491.2
	...	2601.0	2649.6	5529.6
	4387.2	...	2678.4	5568.0
	4502.4	2908.8	2899.2	
	4588.8	3004.8	2908.8	
	4684.8	...	...	
	4780.8	3465.6	3945.6	
	...	3542.4	...	
	6038.4	3792.0	4156.8	
	6076.8	...	4281.6	
	6124.8	3916	...	
		...	4579.2	
		4204.8	...	
		4272.0	4838.4	
		...	4953.6	
		4521.6		
		4675.2		
		...		
		4828.0		
		4924.8		

Table 7.3. Eigenfrequencies of the complete guitar body

Body Frequencies	Frequency Values in the Transient from Table 7.2
...	...
397.97	399.90–401.37
...	...
467.98	450.538–454.1
...	...
1009.92	1020.96–1023.36
...	...
1111.97	1099.97–1107.46
...	...
1265.95	1252.13–1271.52
...	...
2097.94	2093.57–2096.64
...	...
3919.82	3916.7–3919.97
...	...
4245.84	4243.3–4246.56
...	...
4569.79	4569.98–4573.25
...	...
4901.76	4896.67–4899.94
...	

how many modes exist, so Table 7.3 contains only those frequencies that can be allocated. This allocation is almost unequivocal. The lowest frequency of 397.97 Hz, which is by far the strongest in the spectrum, deviates by only approx. 2 Hz from the frequencies of the initial transient. The second frequency, of 467.98 Hz, is about 10 Hz too high, and 1009.92 Hz is about 10 Hz too low. All the higher frequencies match very well.

So the values of the frequencies occurring in the spectra of initial transients can only be ascribed clearly to the guitar body as a whole, and not to individual parts of the instrument. It is also virtually impossible to identify the values of the individual parts, such as top plate, back plate and ribs in the spectrum of the whole body. In fact it would be surprising if this were the case, for the couplings between the individual elements must also alter their eigenvalues (eigenmodes or eigenfrequencies respectively). The notion of several coupled bodies, each with its own eigenvalues and each influencing all the others, must be abandoned, at least in the complicated structure of a guitar. The individual body frequencies are always combination modes. The relatively high frequencies of the combination modes are determined by the choice of the elastic moduli. The high values given above have been assumed throughout, to provide a direct comparison of the individual parts of the guitar, such as between the top plate and the back plate, the ribs, etc. Apart from the relatively large fluctuations of the elastic modulus from wood to wood, as mentioned above, we are interested here in the precise development of the initial transient rather than the exact mode frequencies.

7.5 Development of the Initial Transient

The development of the initial transient can now be described as follows. The "pulling" of the top plate – in this case upwards – triggers a slight swelling of the air. When the top plate returns to its original position the air space adapts correspondingly with the same time delay as before. The ribs start vibrating a little later.

Only after 5 ms do the ribs develop their full vibration. The reason for this is the various types of vibrations to which the ribs are subjected. The initial displacement of the top plate rapidly spreads along it. The wave velocity for the top plate is dispersive, while the time required by the initial displacement of the top plate to reach its boundary can be determined in the model.

Figure 7.10 provides a full view of the guitar (visualized by the MTMS software), revealing how it takes around 1 ms from the time of the initial displacement of the top plate by the string for the first top plate movements to reach the ribs. This corresponds to the results in the diagram of the time series. For the ribs this movement is a 90° coupling, and it thus creates within them a longitudinal movement in their y direction. But owing to the transverse contraction within the ribs, this is also converted into a longitudinal movement in the x direction.

Because of the curvature calculation one part of this longitudinal movement is now converted into a transverse bending wave. This is shown in Fig. 7.9 in the time series for the ribs.

Moreover, the main displacement of the ribs only begins after approx. 5 ms (see Table 7.4). This main movement is caused by the vibration of the air. The relatively long period from the movement starting in the top plate to the displacement of the ribs is explained by the movement of the air space inside the guitar. At the beginning the air is displaced by the top plate in the direction of the top plates normal. This is the z direction for the air. The ribs can, however, only be influenced by the air if the air movement occurs in the x or y directions of the air and thus in the z direction of the ribs. Since the pressure changes in the z direction of the air also have effects in their x/y direction, the air is slowly stimulated to move in its x/y plane. The waves have to build up gradually, until after approx. 5 ms they have reached an order of magnitude such that the air presses on the ribs at the underside of the guitar all together for the first time. Only then does the displacement of the ribs reach a size that continues from that point onwards. The development of the displacement of the enclosed air space purely in the x/y plane can be seen in Fig. 7.11 below.

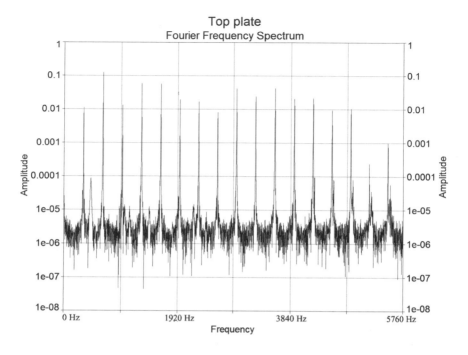

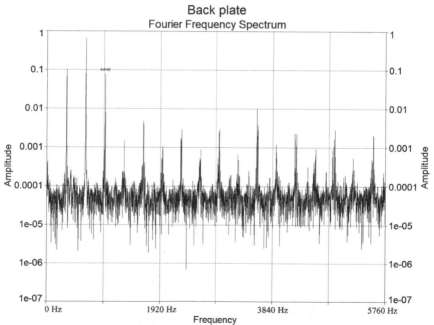

Fig. 7.8. Spectra of the various parts of the guitar from the above diagram, whose time series were integrated over 5 seconds

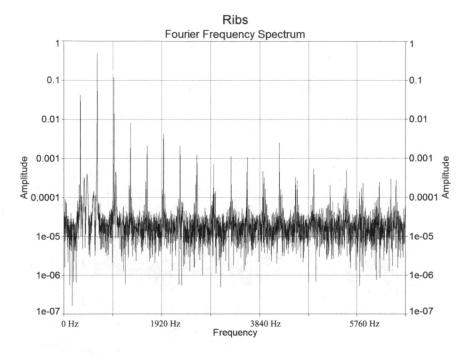

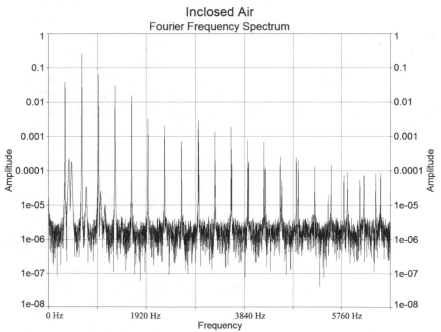

Fig. 7.8 (*continued*)

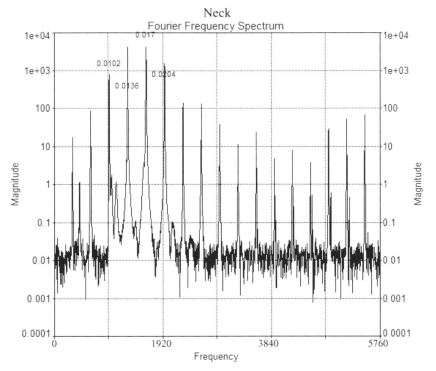

Fig. 7.8 (*continued*)

Since the sound of the ribs has its own characteristics, which are clearly different from the sounds radiated from the other elements of the guitar, this delay in the initial transient has the effect of lengthening its duration. A new nuance is added to the sound after it has started. This "lengthens" the transient and makes it richer in detail. This feature of the ribs is thus an additional part of the initial transient, which makes a decisive contribution to the subtleties of the tone. The sound of the top plate alone, which can be heard in the perception test using the modeled guitar, appears very "simple". It begins immediately and then starts to die away in a linear fashion. The timbre does not contain any more new information, no more "surprises". The quality is strongly reduced. It sounds like a string coupled to just one system of vibration. *The guitar body in its complexity then serves as an enrichment of the initial guitar transient.* Its behaviour just as an resonator would be possible without the guitar body geometry, which is hard to build.

First the back plate is slightly actuated by the ribs, before the air stimulates it to greater vibrations. In the diagram below we can see the back plate movement in the first 50 ms (see Fig. 7.12). The back plate starts to vibrate after 0.2 ms under the influence of the ribs, and after 1.2 ms under that of the air. The air movement, which is shown in the diagram, is integrated over

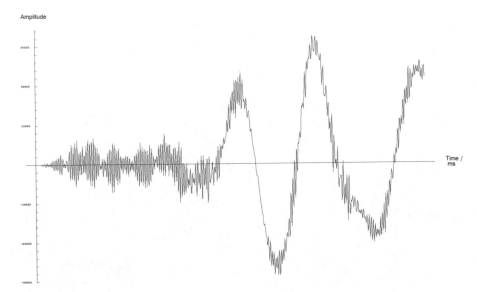

Amplitude

Time /
ms

Fig. 7.9. Development of the initial transient of the ribs for the first 10 ms. Only after 5 ms do the ribs begin to vibrate with a greater amplitude. By this time the vibration of the air, which started at the bridge, has arrived at the ribs and causes transverse movements within them

the layer of air in the x/y plane that is directly below the top plate. Since the height of the ribs is 9.8 cm, the movement requires 0.29 ms to get from the top plate to the back plate if the velocity of a wave in air is 330 m/s. If one imagines the first downward "wave" in the air advancing for approx. 0.3 ms, that is towards the right in the diagram, the minimum of this movement occurs at around 1.2 ms, i.e. when the downward movement of the back plate commences. So we find that the back plate is caused to vibrate 1.2 ms after the string is released, and that these vibrations are large enough to be radiated away and heard.

The neck follows the top plate (see Fig. 7.13). The first displacement of the top plate triggers a response from the neck almost immediately.

This completes the descriptions of the initial transient behavior of all the parts of the guitar. The following table summarizes these descriptions. The time delays from the top plate to the air and the neck can be measured, but they are so small that they hardly influence perception of the sound at all. The time delay for the ribs is given here as 5 ms for both cases: (a) for the attack time after the string is released, and (b) in relation to the top plate, since the beginning of the main attack cannot be determined to ±0.1 ms . But for the perception of the sound this seems to be unimportant.

So, the sequence goes like this: first of all the top plate acts on the air and the neck. The back plate follows after approx. 1.1 ms and only after about 5 ms do the ribs begin to radiate the sound away from the instrument.

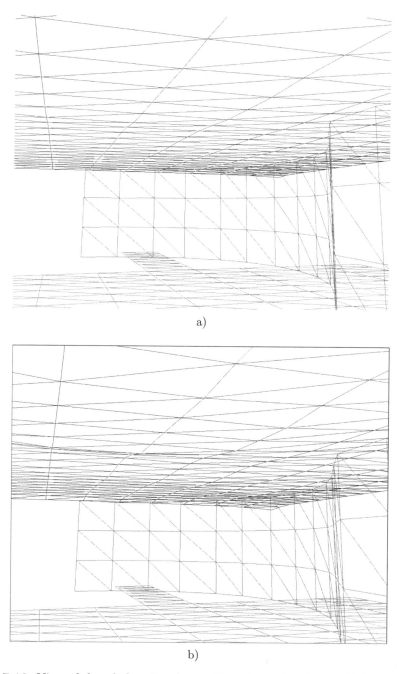

a)

b)

Fig. 7.10. View of the whole guitar from within the enclosed space (guitar part of the MTMS software visualizing a complete view of the guitar). Two points in time are shown: (**a**) Arrival of the string vibration at the top plate, when the top plate begins to move; (**b**) 0.1 ms later, when the first surface waves have reached the ribs

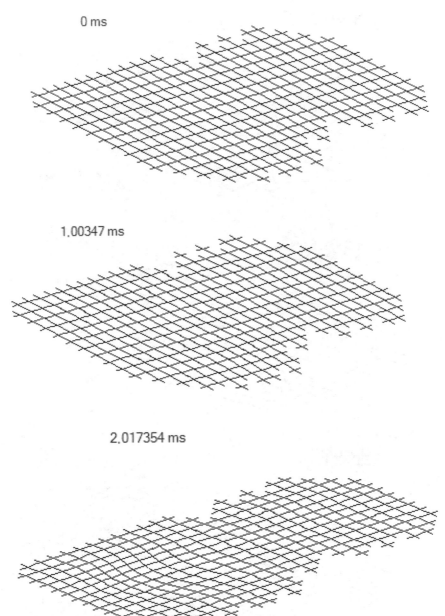

0 ms

1,00347 ms

2,017354 ms

Fig. 7.11. Displacement of the air in the x and y planes for the first 5 ms after onset of the sound. The z influence exerted by the top plate on the air is only slowly transferred into an x/y plane movement of the air, which needs approx. 5 ms before it can all press on the ribs at the underside of the guitar for the first time, thus causing the strong transverse displacement of the ribs. The calculation and visualization were carried out by the guitar part of the MTMS software for the enclosed air space

3,003459 ms

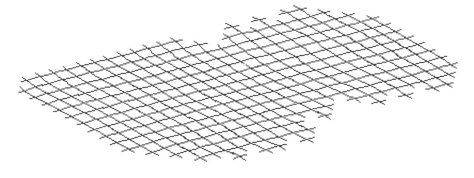

4,010405 ms

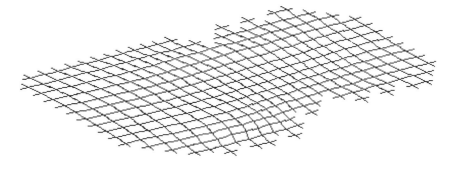

5,000049 ms

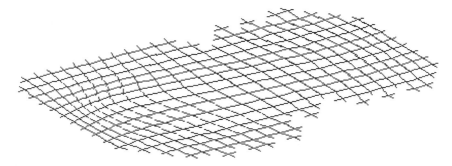

Fig. 7.11 (*continued*)

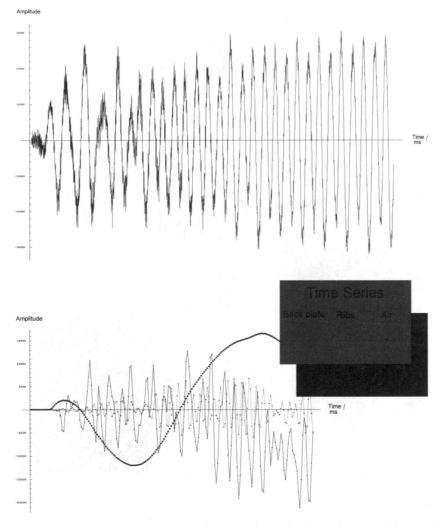

Fig. 7.12. Time series of the radiated transverse back plate vibration (**a**) over the first 50 ms, and (**b**) compared to the ribs and the air over 2 ms. Calculated using the MTMS software

To allocate the individual parts of the guitar body to the overall sound of the instrument it is necessary to gather information about the various timbres. Here the spectral centroids of the radiated sounds are compared with the way the sounds are perceived by human listeners.

Figure 7.14 show the spectral centroids of the top plate, neck, ribs, back plate and air over time (see also Table 7.5). The spectral centroid was calculated from the data of a Wavelet analysis in the range from 100 Hz to 10000 Hz. The plots also show the spectral density.

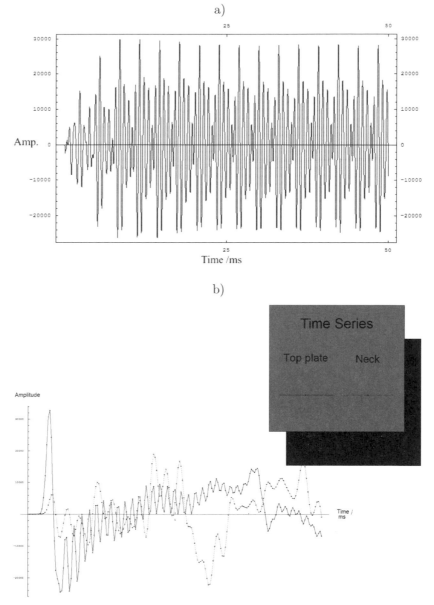

Fig. 7.13. Initial transient of the neck (**a**) over the first 50 ms and (**b**) relative to the top plate during the first 2 ms. Calculated using the MTMS software

The parts of the guitar behave as shown in the sequence of diagrams. The top plate has the highest spectral centroid. It highest value is 2678 Hz, which falls to 502 Hz after 500 ms. The turbulence at the onset of the sound is caused by the initial transient and the fluctuations relate to the amplitude

Table 7.4. Attack times and dependencies of the individual parts of the guitar body after release of the string

Body Part	Attack Time in ms After String is Released	Attack Time in ms After Top Plate
Top plate	0.1	0
Inclosed air	0.135	0.035
Ribs	5	5
Back plate	1.2	1.1 and approx. 0.3 after first major peak for the air
Neck	0.1	0.01

Table 7.5. Spectral centroids of the individual parts of the guitar body for an initial transient. Values given are the maximum value, the minimum value and the average

Body Part	Minimum	Maximum	Average
Top plate	502 Hz	2678 Hz	at $t = 0$ ms: 1500 Hz, at $t = 500$ ms: 502 Hz
Neck	236 Hz	517 Hz	773.5 Hz
Ribs	151 Hz	807 Hz	413 Hz
Back plate	195 Hz	817 Hz	402 Hz
Inclosed air	236 Hz	517 Hz	376 Hz

modulations during the sound. In addition, one can identify the damping of the overtones, which makes the centroid fall logarithmically over time. It should be remembered that at the beginning, the audible center has an average value of around 1500 Hz. If the fundamental tone is 330 Hz for the tone e^1 played on the guitars open high E string, this means that the high frequencies make up an extremely large proportion of the sound. This can also be clearly seen in the Wavelet representation.

The neck has high eigenvalues despite its small geometry compared to the other parts of the guitar. But only during the (quasi) steady state of the sound is it the radiator with the highest spectral value. During the initial transient it is the top plate that dominates, which only relinquishes this prominence after about 300 ms.

The other parts of the body are responsible for the middle frequency range. The air is responsible for the lowest frequencies. Of course all the frequencies are radiated by all the parts, but here we are talking about tendencies allowing the allocation of sound to the most important frequency regions. The ribs and the back plate lie in more or less the same range with very similar minimum and maximum values.

a) Top plate

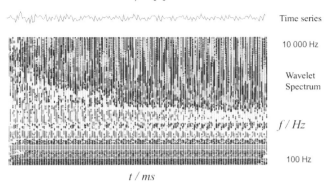

Time series

10 000 Hz

Wavelet
Spectrum

f / Hz

100 Hz

t / ms

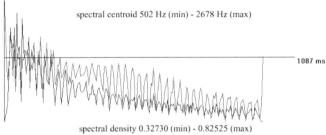

spectral centroid 502 Hz (min) - 2678 Hz (max)

1087 ms

spectral density 0.32730 (min) - 0.82525 (max)

b) Inclosed air

Time series

10 000 Hz

Wavelet
Spectrum

f / Hz

100 Hz

t / ms

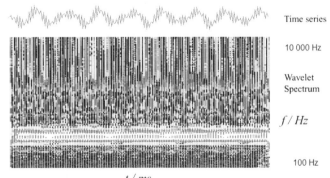

spectral centroid 613 Hz (min) - 934 Hz (max)

652 ms

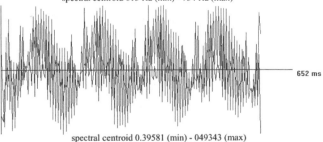

spectral centroid 0.39581 (min) - 049343 (max)

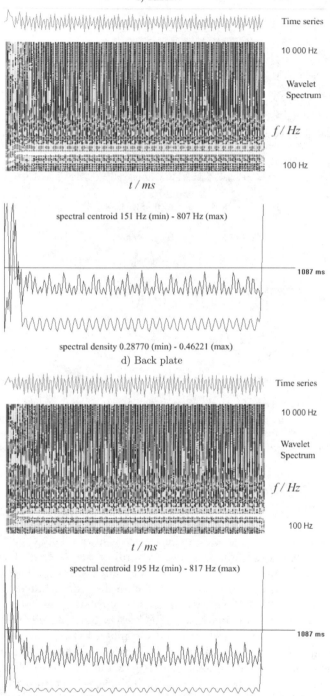

c) Ribs

Time series

10 000 Hz

Wavelet
Spectrum

f / Hz

100 Hz

t / ms

spectral centroid 151 Hz (min) - 807 Hz (max)

1087 ms

spectral density 0.28770 (min) - 0.46221 (max)

d) Back plate

Time series

10 000 Hz

Wavelet
Spectrum

f / Hz

100 Hz

t / ms

spectral centroid 195 Hz (min) - 817 Hz (max)

1087 ms

spectral density 0.29098 (min) - 0.48823 (max)

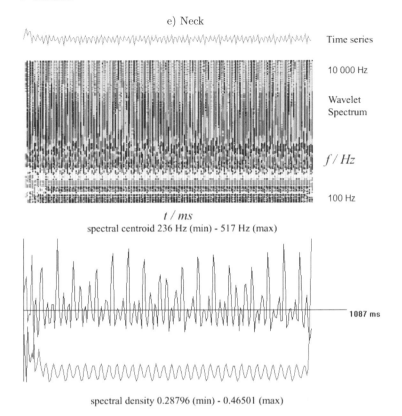

e) Neck

Time series

10 000 Hz

Wavelet
Spectrum

f/Hz

100 Hz

t/ms

spectral centroid 236 Hz (min) - 517 Hz (max)

1087 ms

spectral density 0.28796 (min) - 0.46501 (max)

Fig. 7.14. Wavelet transformations, spectral densities and spectral centroids for the first 500 ms of the sounds radiated by the individual parts of the guitar: (**a**) top plate, (**b**) neck, (**c**) ribs, (**d**) back plate, and (**e**) air. Calculated and visualized using the WAVELET program for time series produced by the MTMS software

7.6 Aural Evaluation of the Calculated Sounds of the Parts of the Guitar

We now come to the perception experiments. No one single part of the instrument produces the familiar sound of a guitar on its own. The top plate reminds one of a harpsichord, being very high and "simple". This means that after the sudden onset of the tone, the sound does not change. The details described above, such as the delayed response of the ribs, are painfully absent here. But it is in the top plate that the attack is at its clearest and most noticeable. The top plate, and only the top plate, is thus clearly responsible for the guitars rapid response and for the brightness at the beginning.

The neck makes its contribution to the sound in the middle frequency range. But it maintains its high frequency values for a certain time $(t > 1\,\mathrm{s})$, in contrast to the top plate. The faster decay in the high frequencies, as is seen in the above diagrams of the spectral centroid, only occurs markedly in the top plate. The other parts of the guitar also experience this decay, but since their centroids are rather small, this only has a small recognizable effect within the first 500 ms. The spectral centroid of the neck, however, fluctuates considerably, which can be explained by the appearance of high frequencies as seen in the Wavelet transformation of this sound. This is probably caused by the high eigenvalues of the neck, which are actuated by approaching low-frequency, but strong, waves in the top plate. In the neck these waves act as impulses that continually stimulate the higher frequencies. So the high eigenvalues are stimulated at regular intervals, and they turn the neck into a radiator of high frequencies.

The air is the part of the guitar that sounds the dullest, reminding one of recordings of classical guitar music from the 1950s and 1960s, for example by Segovia. At that time it was common in the recording studio to have the microphone set up directly in front of the sound hole. So the contribution of the air vibration to the overall sound was large. At the onset of the sound a kind of "pumping" can be heard. This is the sudden and strong onset of the air vibrations, at low frequencies, as we have seen in the above diagrams. The lack of overtones in the air space can also be seen clearly in these pictures as the smoothness of the time series, which produces few overtones.

The back plate and the ribs have the tone that is generally regarded as having "tone color". These are the middle frequencies, which add to the sound – after a time delay, as we have seen. These parts of the guitar give the sound a heaviness owing to their time delay and their sound range. Unlike the sudden onset of the sound from the top plate, the slow swelling of both of these parts is pleasant and gives the sound an initial richness.

If we bring the individual components together, we obtain a guitar tone like one that might be picked up by a microphone positioned 1 m away from the top plate. After the Wavelet transformation of the tone, we get the same properties as those obtained from a tone that has been played on a real guitar and recorded. The Wavelet transformation below in Fig. 7.15 enables us to recognize that, as expected, the string frequencies are disrupted at the beginning by the body resonances. Furthermore, regular amplitude fluctuations can be identified that are produced not only in real guitar tones, but by practically every instrument. The reasons for this have been discussed above in the chapter on the conservation of energy.

The time series of the tone – its first 42 ms – also corresponds to that of a guitars initial transient. The maximum amplitude is reached after an attack time of approx. 6 ms. The eigenvalues of the guitar body still cause individual frequencies to fluctuate markedly during the initial transient. To make these fluctuations easily visible, a greatly increased damping was used for the above diagram, so that the representation is concentrated on the frequency range of the most important disruptions. Here it also becomes clear how one should understand the "struggle" between the string modes and the body resonances. The fundamental periodicity of the time series of the whole guitar tone changes

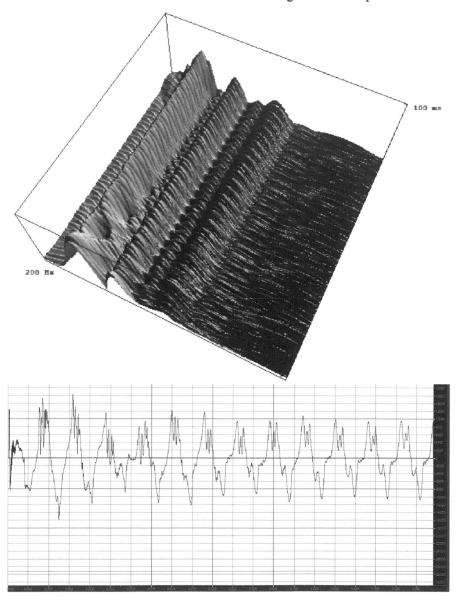

Fig. 7.15. Wavelet transformation of the first 100 ms of a guitar tone with a pitch of e^1 generated using FDM. As with real guitar tones, at the beginning there is a "struggle" between the frequencies of the string and those of individual body resonances. The time series of one of the tones (bottom diagram) shows the typical behavior of an initial transient in the guitar

all the time, fluctuating until it finds its final stationary value. As we saw above, these fluctuations are mainly caused by the presence of the air space, the ribs and the back plate. The top plate itself couples with the string much faster. This "inertia" of the whole body makes the initial transient last longer, so that not only can a simple attack be heard, but the sound develops slowly, and therefore sounds considerably more differentiated.

7.7 Coupling of the String to the Top Plate

In the section about studies of the coupling between the string and the body, I pointed out that theoretical models, including FEM models (Gough 1981), have problems with the simulation of high frequencies. This aspect will now be discussed.

The models used to date made an assumption about the impedance – that the coupling of the string to the body took place through changes in the angle that occurs during vibration. This change is a derivative of the displacement of y from x of the first order $\frac{dy}{dx}$ and is therefore termed an impedance theory of the first order here.

In FDM the impedance of the string to the body is likewise calculated based on the assumption that the strings energy is transferred to the body by the changing angle of the string. This changing angle means that a force is acting on the guitar body. The element to which the string is coupled experiences this force in the z direction of the top plate. The acceleration of this top plate element resulting from the force causes its change in movement and thus the spread of the string movement to the top plate and finally to the whole guitar. The advantage of this line of consideration is that no theoretical assumptions are necessary about how the top plate picks up the frequencies. The geometry and the notion of change in the angle alone lead to the result.

It was found that the simple angle change is insufficient to simulate the guitars richness in overtones. The simplified impedance explanation appears not to confirm the more complicated geometric ratios of the top plate sufficiently. Probably it is the assumed single point transfer that is insufficient. To achieve realistic results a second term must be added to the impedance calculation of the first order, a term of the second derivative.

Figure 7.16 shows the development of the time series for the impedance calculation of the first and second orders. The difference is obvious. While the first order calculation only transfers components with a low frequency, the second order calculation provides the components with high frequencies. This is easily explained if the time series are compared as they arise at the bridge.

Figure 7.17 compares two time series. Both of them were made after the guitars high E string was plucked at the bridge. The top time series shows the angle change, while the bottom series represents the second spatial derivation, that is, the curvature of the string at the location of the bridge. It can clearly be seen that both time series contain both the fundamental period of the string

Amplitude

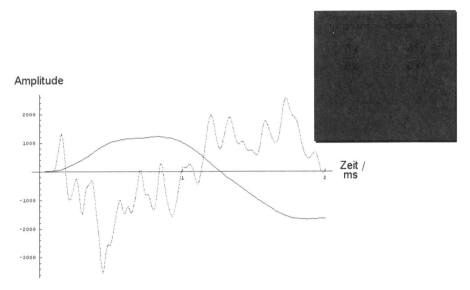

Fig. 7.16. Comparison of time series of the top plate for the cases of impedance calculation of the first and second order

and an overtone structure. The rapid movement, which is superimposed on the coarse one, arises because of the inertia of the parts of the string that are not involved in the large displacement, and because of the damping. The main point of plucking, which exists at a point at the beginning, becomes rounder with time, owing to the damping (also compare Fig. 7.5 in this chapter; the string is divided into two parts due to its trapezoid shape; these parts vibrate also their microstructure). The reflection at the ends is not instantaneous, but takes a certain amount of time. This allows energy to wander to and from the point of maximum displacement and the fixed points of the string, here the bridge and the saddle. These small energy reflections generate the components with high frequencies. Since these components repeat themselves with the large periodicity, they represent an additional boost to the overtone

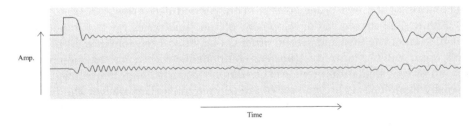

Fig. 7.17. Time series of the tone e^1 on the high E string after plucking at the bridge. *Top*: with a change to the angle. *Bottom*: with the second derivative of the displacement

structure of the string. This is because the actual overtone content of the vibration is normally obtained from the basic form of the plucking. This form is not sinusoidal and can therefore be broken down into Fourier components, which must all be harmonics of the fundamental frequency, since they all have the same fundamental period. De facto, for the point on the string in contact with the bridge, the angle change over time is to be viewed as a periodic impulse. This impulse must be taken up by the sluggish top plate. Now, the impulse is long and strong enough to initially "lift" the top plate with quite a force. Afterwards the string leaves the top plate in peace until the next lift. As the lift is periodic, the fundamental frequency is very pronounced. And because we are considering an impulse, if it is short enough, the eigenvalues of the guitar body are also stimulated. But it is very difficult for the overtones to be generated. This is not because they are not available, but because the transfer of the strings angle change affects a medium that occupies a region in space – the stiff top plate.

The situation changes fundamentally if the second derivative of the strings vibration is included in the calculation. As shown in Fig. 7.17, the strings vibration is a second spatial derivation and not so very strongly influenced by the large "boosts" coming from the string, but rather by the smaller displacements repeating periodically. The overtone structure is considerably reinforced. This actuation alone would, however, make the sound much too bright, so it would no longer sound anything like a normal guitar tone. It is only the combination of the two types of impedance, that of the first order and that of the second, that produces a realistic guitar tone.

The physical reason for this is also easy to see. The differential equation of the strings vibration, as discussed above, gives the curvature of the string at the point considered to be the cause of the restoring force for each and every string element.

$$-F_{\text{acceleration}} = F_{\text{backdriving}} = k \, \frac{d^2 y}{dx^2} \tag{7.1}$$

There is no reason not to assume that this also applies to the relevant point on the bridge. So in a physical view, the curvature of the string at this point should also be taken into account, and if the curvature is left out the reasons for this should be explained

7.8 Pre-Scratch

"Pre-scratch" is a term used here to refer to the sound component that precedes the actual tone. It is caused by the finger rubbing along the string before it releases it. In the apoyando and tirandu techniques the finger is normally placed on the string in such a way that both the fingernail and the fingertip touch the string. The plucking action is actually the closure of one or more fingers to form a fist. In fast playing, the fingers generally do not touch the palm of the hand and the movement of the fingers remains quite small. Now, since the position of the fingers is usually not parallel, but instead at a certain angle to the strings, the fingers must slide along the strings slightly, before releasing them.

To produce a forte on the classical guitar, the starting position of the finger on the string is not relaxed. The finger first presses the string downwards and remains there until the tone is played. For a loud tone the displacement of the string can exceed one centimeter. The larger the initial displacement, the louder the tone. For this reason, the displacement must be effected in the direction of the guitars normal – "into the guitar", as a guitarist would say – since the resulting transverse string vibration will then run perpendicular to the top plate and thus perpendicular to the bridge. It is this movement that causes a force that changes over time perpendicular to the top plate and that then feeds into a transverse movement of the top plate. This is important because only this transverse displacement has sufficient radiating surface to be loud enough to be heard by the listeners. There also exists a transverse movement of the string parallel to the top plate that causes only a slight tilting of the bridge, which is so small that the energy radiated by it would not generate a guitar tone. The transverse movement of the string parallel to the top plate can only have any audible influence on the sound if there is a coupling of longitudinal and transverse vibrations, or of these types of vibrations with bending vibrations in the string. In such cases there would be a transfer of energy from one type of vibration to the next. The transverse movement parallel to the top plate would influence either the bending wave, the torsion wave or the longitudinal movement, which would then in turn exert an influence on the transverse movement of the string at right angles to the top plate. We would have a situation similar to that with the coupling of the top plate to the back plate via the ribs. The radiating bending wave in the top plate is converted into a non-radiating longitudinal wave in the ribs, which then turns back into a radiating movement of the back plate. The sound wanders through the guitar in what we could call a "subterranean" fashion. For the string, this would correspond to the types of vibrations that influence the transverse movement of the string perpendicular to the top plate and thus the sound of the guitar, not directly – but indirectly via coupling.

But here we are interested in the finger noise. The cause of this noise can be investigated thus: to test the guitar model the sound is generated and compared with the real finger noise. In earlier studies (Bader 2002a) components of the later harmonic overtone structure of the string were found in the finger noise. The noise is of very short duration, lasting somewhere between 1 ms and 5 ms. There is a clearly audible difference depending on whether the noise occurs on a treble string or a bass string. The three bass strings are made of nylon spun with steel. The steel binding is necessary because otherwise the strings intended to produce tones of the desired pitches at an acceptable tension would be so thick that they would experience very severe damping and their own stiffness would interfere. So binding the strings with steel helps to make them playable.

The development of the pre-scratch is assumed to be as follows. If the finger and the fingernail slide along the string, their movement is described by the friction of the finger in contact with the string. This friction can be either sliding friction or static friction. As the finger moves forwards, sliding friction must be involved. But if the finger were to glide over the string with sliding friction only, there would be no reason for any tone. The noise that is generated would simply be explained by alternation between sliding friction and static friction. But this is what happens, in detail: at the beginning of the movement the finger is at rest and there is static friction between it and the string. The muscles exert a force that sets the finger in motion towards the palm of the hand, and this results in a force within the finger parallel to the string. Since the fingertip is soft, the skin is initially somewhat stretched, but the finger still sticks to the string. Only if the force exceeds a certain value does the finger "escape" from the string and move along the string a little. This is the temporal area of the sliding friction, because the finger does not slide along the string with zero resistance. But since sliding friction is much weaker than static friction, the finger can glide along the string a little way almost effortlessly. But the mobility of the skin and of the fingertip is much greater, and therefore the movement is much faster than that of the whole finger. The fingertip thus reaches a point on the string where the finger would be, if it had not been held in position at the beginning of the movement. Maybe we can also assume that the skin of the fingertip moves a bit past its equilibrium position in relation to the finger, for a brief moment, and is then rapidly slowed down again. In both cases it is again brought to a standstill and then experiences static friction again. The whole process starts again from the beginning.

This description applies to the treble strings. For the bass strings there is the additional factor of the ribbed surface of the strings because they are spun with steel. It can be assumed that the finger "gets stuck" in each groove of the surface. This effect is often used by rock guitarists to create a longer-lasting noise. The plectrum that is used contacts the string at right angles and is then drawn along the string. The result is an effect creating a certain tonality that depends on the velocity with which the plectrum travels along the string. This can be easily understood, for the velocity of the movement determines how often per second the plectrum catches in a groove. This results in a periodicity, which must be recognizable as a pitch. The fact that here no sine-wave tones are generated is due to the impulse nature of the generator. Every time the plectrum "sticks" in one of the grooves it gives rise to a small impulse and not a sine wave.

Figure 7.18 shows the spectrum of the air vibration that forms part of the finger noise. These sounds were achieved by manipulations in the string part of the MTMS software. Here the presence of the individual string modes can clearly be seen. In the diagram the first six modes that are summarized in the table below are indicated by the diagonal lines. The other modes of the string are, however, also clearly visible.

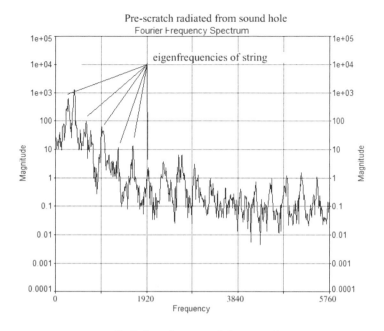

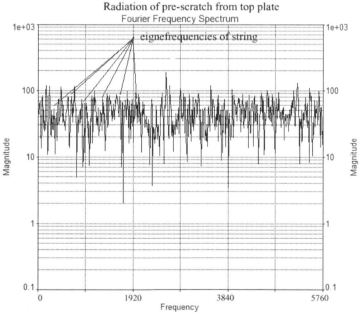

Fig. 7.18. Spectra of the pre-scratch for the top plate and enclosed air space. The eigenvalues of the string can be clearly recognized in the air spectrum. In the spectrum of the top plate they are almost masked by the large number of actuated modes. The sound of the air is approximately periodic, while the sound of the top plate is made up of noise

Table 7.6. Spectral string components in the spectra of the back plate, the top plate and the air for the finger noise, given in Hz. Calculation using manipulations in the string part of the MTMS software

Mode	Top Plate	Back Plate	Inclosed Air
1	326.4	336.0	336.0
2	681.6	652.8	652.8
3	969.6	988.8	988.8
4	1305.6	1315.2	1315.2
5	1651.2	1632.0	1632.0
6	1987.2	1987.2	1948.8

Table 7.7. Frequency ratios for the individual string modes of the back plate, the top plate and air for the finger noise. The back plate and the air space are also in the mode ratio for the fundamental mode, which is no longer $f_0 = 336$ Hz, but $f_0 = 326.4$ Hz (for reasons, see Table 7.6)

Mode	Top Plate	Back Plate	Back Plate with $f_0 = 326.4$ Hz	Inclosed Air	Inclosed Air with $f_0 = 326.4$ Hz
1	1.0	1.0	1.0	1.0	1.0
2	2.088	1.943	2.000	1.943	2.000
3	2.971	2.943	3.029	2.943	3.029
4	4.000	3.914	4.029	3.914	4.029
5	5.059	4.857	5.000	4.857	5.000
6	6.088	5.914	6.088	5.800	5.971

Table 7.6 contains the frequencies of the individual string modes for the three guitar parts top plate, back plate and air space. Table 7.7 contains the frequency ratios of these modes to their relevant fundamental modes.

The ratios of the modes are very good for the top plate, the back plate and the enclosed air space, assuming that the fundamental tone found for the top plate $f_0 = 336$ Hz. The shift of the fundamental tone can also be identified in experimental data (Bader 2002a).

The finger noise was taken to be produced by a fixed period of sliding of the fingertip along the string, which is, however, subject to stochastic fluctuations. Here a fundamental period of $\Delta T = 0.21$ ms was selected for ΔT with a fluctuation of $\Delta t = .104$ ms. In preliminary studies in which the stochastic fluctuation was not included in the calculation, no realistic finger noise could be generated. Since ΔT was strongly periodic there, it generated a pitch that could be clearly identified, $f_0 = 4800$ Hz, but not the pre-scratch.

So if one agrees with the theory of the shift between static friction and sliding friction, one must assume that the "escape" has a periodicity, but that this periodicity is not constant. The fact that it cannot be absolutely constant is shown by a simple attempt to slide a finger along the string. The velocity of the slide determines the brightness and a type of tonality of the "sliding tone", which is the finger noise itself. But there remains a noisy sound, although the velocity naturally also raises the necessary period of escape $\triangle T$. However, this escape period still remains stochastic – otherwise no noise would arise, but instead a high-pitched tone.

8

Summary and Outlook

In this monograph I have shown that the vibrating behavior of the guitar can only be described correctly if the analysis includes the transient aspect. The new paradigm of transient instrumental acoustics has proved its worth in many areas and produced astonishing results.

1. All of the guitars modes – not only the three lowest – are combination modes. The eigenfrequencies occurring in the assembled instrument do not correspond to the eigenfrequencies of the separate parts of the guitar in isolation. Obviously this applies to the top plate and the back plate before they are glued to the rest of the guitar, because in their free state their frequency modes occur in the presence of free boundary conditions and are thus completely different. I have shown that components of the guitar which are under tension but de-coupled from all the other parts, such as the top plate, the back plate and the ribs, vibrate at their own frequencies, and these frequencies no longer occur when there is active coupling. It must be concluded that a geometry such as this only possesses combination modes. It is very likely that this finding is transferable to all instruments. Any analysis of the modes that considers the parts of the guitar as separate entities will therefore only be of use as a guideline reflecting the approximate spectrum of frequencies. The problems that violin-makers, for example, have in getting a reliable impression of what the whole violin will sound like later, by knocking on separate top plates and back plates, speak volumes.

2. The coupling of the strings to the top plate is also totally different than what has been supposed to date. It is clearly not possible to assume an impedance calculation of the first order for the transfer of the vibrations of the strings onto the body. The transfer must take account of the second derivative. This problem was only brought to light by the results of experimental measurements (Woodhouse 2003b). The reason for the delay in recognizing that previous suppositions about the type of coupling do not work lies in the experimental investigation methods used to date. They

assumed pure resonator couplings. But during the transient, the body of the guitar does not simply act as a resonator. The agreement of earlier experimental results concerning the coupling with the previous calculations was simply due to the fact that a guitar string had never been plucked in the laboratory. The coupling was generated at one point only by the transfer of energy from an actuator. In such a case there is no coupling between the string and the top plate, but instead a coupling between the actuator and the top plate. In mathematical terms this is a different situation altogether. The cause of the discrepancy between the theory to date and the results of measurements using strings that are genuinely plucked was revealed by the iterative procedure adopted in this study, and a solution was found. The new theory is also physically plausible. Since the correct type of coupling is nonlinear, additional effects may be expected here, with the result that suitable adaptations to the construction of the guitar will be able to elicit marked changes in the sound.

3. This study has also discovered the reason why guitars are built with ribs – and not with round bodies, like lutes. The transient behavior of the interior air space is very complex, since the geometry is three dimensional. In this case an analytical solution is no longer possible, and therefore only the iterative guitar calculation proposed here was able to reveal the behavior. The influence of the ribs via this interior air space is subject to a considerable time-delay owing to the complicated couplings between the three air space directions in the transient behavior. This leads to a kind of "echo" with the typical timbre of the ribs. This part of the sound, which is added to the initial transient, makes the guitars tone more complex and thus richer and more refined than would be the case if the string were fixed only to a simple board. The "echo" in the initial transient is very typical of guitars. It makes the tone heavier and subjectively louder. This improves the characteristics of the guitar for solo performances. The effect fully justifies the considerably more complicated construction of a guitar *with* ribs.

4. The sounds from the individual parts of the guitar exhibit a clear pattern of separate effects contributing to the overall sound of the guitar. The general view held until now, that the sound of a guitar is mainly radiated away from the instrument by the top plate, is incorrect. The top plate produces a very bright tone that commences immediately and is responsible for the guitars attack. But on its own it sounds boring. It lacks the time-delay caused by the ribs, along with the deep frequencies. This finding also agrees with the statements of guitar-makers with whom I have discussed this problem. They confirm that the top plate is only responsible for the fast onset of the tone. The deep frequencies are produced by the air space, which radiates the sound out through the sound hole. If one listens to the tone produced by the MTMS software only taking account of the sound radiated by the air space, one is reminded of classical recordings of guitars made in the 1950s and 1960s (e.g. of Segovia). At that time the microphone was positioned very close to the sound hole because it was thought that the true tone of

the guitar would be found at this location. This is what caused the hollow, pumping sound heard on these recordings.

The back plate and the ribs, in their turn, share the middle range of frequencies, the back plate being responsible for the deeper frequencies and the ribs responsible for the upper middle range. The time-delay of the ribs has already been discussed under point (6.3) above. The back plate begins to vibrate with hardly any delay at all, but it lacks completely the attack found in the top plate. This corresponds to the range of frequencies radiated by the back plate, if one wishes to associate a hard attack with brightness and a softer attack more with a dull sound. One guitar-maker I spoke to told me that the sound of his guitars was generated by the back plate, and not by the top plate. What he meant was the fact that we do not associate "sound" in the musical sense solely with brightness. The brightness contributes to the frequency range of the top plate. But the back plate produces and radiates those frequencies that are in the musically variable range within which the sound can be finely tuned, as every sound engineer knows. At this point we should remember that here we are dealing with radiated frequencies, and not necessarily the eigenvalues of the parts of the guitar. Only a transient analysis of the guitar system can determine the vibrating behavior of the parts of the guitar; a static analysis can never do this.

5. The pre-scratch is highly significant when it comes to identifying the instrument as a guitar. It, too, already contains in a rudimentary form the spectrum of the tone about to be played. The production of the noise is considerably more complicated than was initially supposed. Damping must occur at the point of contact between the finger and the string. However, it does not damp the string completely, and is slight enough for a tone to be produced. And apart from this, the distribution of the time periods in which the finger slips and catches again must be stochastic, because otherwise the tone dependent on the speed of the finger would be the only one that was audible. In that case it would be impossible to hear the frequencies of the string amongst all the other tones, and there would not be any noise. This means that the finger noise can be characterized by three parts: the white noise, the tone dependent on the speed of the finger, and finally the eigenfrequencies of the string. Since the velocity of the finger changes constantly, this tone also changes. With some finger noises a pitch-glide occurs from the deeper frequencies to the higher ones. This effect forms the basis of the musical event, associated with a tense expectation – which is then followed by the actual sound of the guitar.

6. When the back plate of the guitar is under tension it exhibits both an increase in its eigenfrequencies and an overall increase in brightness, compared to a back plate in the relaxed state. Guitar-makers name this as the reason for putting it under tension. We find that a geometry with an initial tension has higher eigenvalues, even though in the case of initial tension the vibrations also occur around a state of equilibrium where no forces

appear. The reason for this is the coupling between transverse and longitudinal movements of the plate. The fact that this coupling must occur is explained by the tension of the guitars back plate in the longitudinal direction, although the bulge itself is transverse. So there must be a coupling between transverse and longitudinal forces. The coupling proposed here results in the behavior observed in guitars. The longitudinal squeezing causes the back plate to distend. When the vibrations occur, the transverse movement does not only have to overcome the rigidity, but also has to deal with the longitudinal squeezing. The result is an increase in the eigenfrequencies and a generally enhanced brightness of the sound, heard when guitars are played.

7. Coupling between longitudinal and transverse waves was proven and calculated for the curved back plate, both theoretically and iteratively (see (6) above). In addition it was possible to illustrate and/or suggest a theory of the spread of the waves and coupling of the two types of waves in curved structures, along with effects that occur in flat plates. The curvature of the ribs results in a continual transformation of longitudinal waves into transverse waves through the constant discontinuity of the curvature. This, too, raises the eigenvalues of the ribs. The extreme forms of this transformation are found in the coupling between the top plate and the ribs, and in that between the back plate and the ribs. At these points the curvature is 90°, so the transverse waves are completely converted into longitudinal waves, and vice versa. For example, the transverse waves in the top plate are transformed at the edge of the top plate into longitudinal movements of the ribs. Since they cannot be radiated away, we can say that they reach the back plate via a "subterranean route". Upon arrival at the back plate they are turned back into transverse waves because of the 90° coupling. Such complicated sequences of movements cause constant fluctuations even in the quasi-stationary part of the guitars sound, which are quite familiar to us from real guitar tones. Furthermore, a combination of longitudinal and transverse waves in a flat plate can lead to changes in the eigenvalues in certain circumstances, especially if the displacements are large. The peak of the spectral centroid at the beginning of the top plates transient (see chapter on results) can be explained in this way. It is also possible that synergetic phenomena occur here, if the displacements are large enough, because in this case we are dealing with a system of two nonlinear, coupled and very complex differential equations.

8. The iteration of the string by FDM, which was only calculated in this study for the transverse situation, reveals fine structures within the sound that are observed in real strings. For example, the initially trapezoidal displacement of the string (caused by the finger) moves along the string. The initially clear peak of the displaced string becomes more and more indistinct, owing to the damping of the string. In addition, the parts of the string to the right and left of the peak vibrate at their own eigenvalues with a smaller amplitude, while the main peak wanders about. This results in

higher parts to the spectrum whose frequency values are constantly changing, since the "wandering" of the peak means that the partial vibrating lengths of the string are likewise constantly changing. So there are many frequency components in the high range of the spectrum, but they are very quiet. At the same time these components are always coupled to the top plate and thus add further richness. It is extremely difficult to describe this behavior analytically. The FDM of the MTMS software allows these additional vibrations to be observed in slow-motion.

The results from iterative instrumental acoustics are very encouraging. This is because, starting from the interesting physical results from instrumental acoustics, they impinge on issues occurring in the study of music. Both here and in the area of nonlinear dynamics there are many unanswered questions about the behavior of musical instruments, many of which are interrelated and have not even been looked at. The iterative algorithms and software packages presented here can reveal the behavior of musical instruments in ways that would have been inconceivable with the methods used previously. In future a broad field of research will arise here, which will close the gap between physics and musicology.

References

Argyris, J.: *Die Erforschung des Chaos: Studienbuch für Naturwissenschaftler und Ingenieure.* Vieweg, Braunschweig, 1995.

Backus, J.: *Multiphonic tones in the woodwind instruments.* In: Journal of the Acoustical Society of America, 63, 591–599, 1978.

Bader, R.: *Physical model of a complete classical guitar body.* In: Proceedings of the Stockholm Music Acoustics Conference 2003, R. Bresin (ed.) Vol. 1, 121–124, 2003.

Bader, R.: *Fraktale Dimensionen, Infomationsstrukturen und Mikrorhythmik der Einschwingvorgänge von Musikinstrumenten.* PhD der Universität Hamburg. 2002.

Bader, R.: *Fractal correlation dimensions and discrete-pseudo-phase-plots of percussion instruments in relation to cultural world view.* In: Ingenierias, Octubre-Diciembre 2002, Vol. V. No 17, p. 1–11.

Bader, R.: *Fractal dimensions of initial transients of musical instruments.* In: Journal of the Acoustical Society of America, Vol 110, 5, 2755, 2001.

Bathe, K.-J.: *Finite-Elemente-Methoden.* Springer 2002.

Bank, B. & Sujbert, L.: *Modeling the longitudinal vibration of piano strings.* In: Proceedings of the Stockholm Music Acoustics Conference 2003, S. 143–146, 2003.

Bensoam, J., Causse, R., Vergez, C., Misdariis, N. & Ellis, N.: *Sound Synthesis for Three-Dimensional Objects: Dynamic Contact between two Arbitrary Elastic Bodies.* In: Proceedings of the Stockholm Music Acoustics Conference 2003, S. 369–372, 2003.

Berry, F.A., Herzel, H., Tieze, I.R. & Story, B.H.: *Bifurcations in Excised Larynx Experiments.* In: Journal of Voice 10, 129–38, 1996.

Beurmann, A. & Schneider, A.: Sonological Analyses of Harpsichord Sounds. In: Proceedings of the Stockholm Music Acoustics Conference S. 167–170, 2003.

Bohlen, Th.: *Viskoelastische FD-modellierung seismischer Wellen zur Interpretation gemessener Seismogramme.* PhD, Kiel 1998.

174   References

Borg, I.: *Entwicklung von akustischen Optimierungsverfahren für Stabspiele und Membraninstrumente. (Development of acoustical optimization methods for musical bars and membrane instruments).* PTB Report, Projekt 5267, Braumschweig 1983.

Boullosa, R.R.: *The use of transient excitation for guitar frequency response testing.* In: catgut Acoustical Society Newsletter, 36, 17, 1981.

Bretos, J., Santamaria, C. & Moral, J.A.: *Vibrational patterns and frequency responses of the free plates and box of a violin obtained by finite element analysis.* In: Journal of the Acoustical Society of America, 105,3, 1942–1950, 1999.

Bretos, C., Santamaria, C. & Moral, J.A.: *Finite element analysis and experimantal measurements of natural eigenmodes and random responses of wooden bars used in musical instruments.* In: Applied Acoustics 56,3, 141–156, 1999.

Caldersmith, G.: *Guitar as a reflex enclosure.* In: Journal of the Acoustical Society of America 63, 1566–1575, 1978.

Caldersmith, G.: *Frequency response and played tones of guitars.* Quarterly Reports STL-QPSR 4/1980, Department of Speech Technology and Music Acoustics, Royal Institute of Technology KTH Stockholm, 50–61, 1980.

Caldersmith, G.: *Towards a classic guitar family.* In: American Lutherie, 18, 20–25, 1989.

Caldersmith, G.: *The guitar family, continued.* In: American Lutherie, 41, 10–16, 1995.

Caldersmith, G.: *Designing a guitar family.* In: Applied Acoustics 465, 3–17.

Christensen, O. & Vistisen, B.B.: *The response of played guitars at middle frequencies.* In: Acustica 53, 45, 1983.

Christensen, O.: *Guitar Sound Pressure Response.* In: Acustica 54, 289–95, 1984.

Cook, P.R.: *Nonlinearity, Waveshaping, FM.* In: Real Sound Synthesis for Interactive Applications. Cook (ed.) AK Peters, Natick, 109–120, 2002.

Cremer, L.: *Der Einfluß des Bodgendrucks auf die selbsterregten Schwingungen der gestrichenen Saite.* In: Acustica 30, 119, 1974.

Cremer, L: *The physics of the violin.* MIT Press, Cambridge, 1985.

David, M.B.: *Charactérisation acoustigues de structures vibrantes par mise en atmosphére raréfié.* PhD Thesis de l'Université Paris 6, 1999.

Derveaux, G.: *Modélisation numérique de la guitare acoustique.* PhD Thesis de l'Ecole Polytechnique Paris, 2002.

Elejabarrieta, M.J., Ezcurra, A. and Santamaria, C.: *Vibrational behaviour of the guitar soundboard analysed by means of finite element analysis.* In: Acustica united with Acta Acustica, 87, 128–136, 2001.

Elejabarrieta, M.J., Ezcurra, A. and Santamaria, C.: *Coupled modes of the resonance box of the guitar.* In: Journal of the Acoustical Society of America, 111, 2283–2292, 2002.

Feynman, R.P.: *Vorlesungen über Physik. Vol II.* The Feynman Lectures on Physics, Vol II, in deutscher Bearbeitung, Oldenbourg Wissenschaftsverlag GmbH, 2001, 1963.

Fitch, T., Neubauer, J. & Herzel, H.: *Calls out of Chaos: The Adaptive Significance of Nonlinear Phenomena in Mammalian Vocal Production.* In: Animal Behavior 63, 3, 407–18, 2002.

Fletcher, N.H.: *Mode locking in nonlinearly excited inharmonic musical oscillators.* In: Journal of the Acoustical Society of America 64, 1566–69, 1978.

Fletcher, N.H.: *Nonlinear frequency shifts in quasi-spherical cap shells: Pitch glide in Chinese gongs.* In: Journal of the Acoustical Society of America, 2078, 2069–2073, 1985.

Fletcher, N.H. & Rossing, Th.D.: *The Physics of Musical Instruments.* Springer 2000.

Flügge, W.: *Statik und Dynamik der Schalen.* Berlin/Göttingen/Heidelberg 1962.

Galluzzo, P.M. & Woodhouse, J.: *Experiments with an automatic bowing machine.* In: Proceedings of the Stockholm Musical Acoustics Conference 2003, S. 55–58, 2003.

Gear, C.W.: *Numerical Initial Value Problems in Ordinary Differential Equations.* Prentice-Hall, Englewood Cliffs, NJ, 1971.

Gibiat, V.: *Phase space representations of acoustical musical signals.* In: Journal of Sound and Vibration, 123/3, 529–536, 1988.

Gibiat, V. & Castellengo, M.: *Period Doubling Occurences in Wind Instruments Musical Performance.* In: Acustica 86, 746–754, 2000.

Gough, C.E.: *The theory of string resonances on musical instruments.* In: Acustica 49, 124–141, 1981.

Goudreau, G.L. and Taylor, R.L.: *Evaluation of Numerical Integration Methods in Elastodynamics.* In: Computer Methods in Applied Mechanics and Engineering, 2, 69–97, 1972.

Güth, W.: *"Ansprache" von Streichinstumenten.* In: Asustica, 46, 259–267, 1980.

Güth, W.: *Einführung in die Akustik der Streichinstrumente.* Stuttgart 1995.

Güttler, K.: *Wave analysis of a string bowed to anomalous low frequencies.* In: Journal of the Catgut Acoustical Society 2, 8–14, 1994.

Haase, M., Widjajakusuma, J. & Bader, R.: *Scaling Laws and Frequency Decomposition from Wavelet Transform Maxima Lines and Ridges.* In: Novak, M.M. ed.: Emergent Nature. World Scietific 2002, p 365–74.

Haile, J.M.: *Molecular Dynamics Simulation. Elementary Methods.* John Wiley & Sons, Inc., NY 1997.

Haines, D.W.: *On musical instrument wood.* In: Catgut Acoustical Society Newsletter 31, 23–32, 1979.

Haken, H.: *Synergetik.* 3. Auflage, Springer, 1990.

Haken, H.: *Advanced Synergetics.* Springer 1983.

Hanson, R.J., Schneider, A.J. & Halgedahl, F.W.: *Anomalous low-pitched tones from a bowed violin string.* In: Journal of the Catgut Acoustical Society 2, 1–7, 1994.

Hanson, R.J., Macomber, H.K. & Morrison, A.C.: *Unusual motions of a nonlinear asymmetrical vibrating string.* In: Proceedings of the Stockholm Musical Acoustics Conference 2003, 731–34, 2003.

Hartmann, W.M.: *Signals, Sound, and Sensation.* Springer 1998.

Helmholtz, H. von: *Die Lehre von den Tonempfindungen.* 1862. Hier in der 7. Auflage von 1968.

Hilber, H.M.: *Analysis and Design of Numerical Integration Methods in Structural Dynamics.* In: EERC Report No. 77–29, Earthquake Engineering Research Center, University of California, Berkeley, California, November 1976.

Hilber, H.M., Hughes, T.J.R. and Taylor, R.L.: *Improved Numercal Dissipation for Time Integration Algorithms in Structural Dynamics.* In: Earthquake Engineering and Structural Dynamics, 5, 283-292, 1977.

Holzapfel, G. A.: *Nonlinear Solid Mechanics. A continuum approach for engineering.* Wiley 2000.

Houtsma, A.J.M. & Burns, E.M.: *Temporal and spectral characteristics of tambura tones.* In: Journal of the Acoustical Society of America 71, S83, 1982.

Hughes, J.R.: *The Finite Element Method. Linear Static and Dynamic Finite Element Analysis.* Dover Publications, Mineola, 1987.

Hutchings, C. (ed.): *Research Papers in Violin Acoustics.* Band I and II, Publication by the Acoustical Society of America, 1997.

Jahnel, F: *Die Gitarre und ihr Bau.* Frankfurt a.M. 1986.

Jahnke-Ende: *Tafeln höherer Funktionen.* Leipzig 1948.

Jansson, E.V.: *A study of acoustical and hologram interferometric measurements of the top plate vibrations of a guitar.* In: Acustica, 25, 95–100, 1971.

Jansson, E.V.: *Fundamentals of the guitar tone.* In: Journal of Guitar Acoustics, 6, 26–41, 1982.

Jansson, E.V.: *Acoustics for the guitar player.* In: Function, Construction, and Quality of the Guitar. Jannson (ed.). Royal Swedish Academy of Music, Stockholm, 7–26, 1983.

Karjalainen, M.: *Time-domain physical modeling and real-time synthesis using mixed modeling paradigms.* In: Proceedings of the Stockholm Musical Acoustics Conference 2003, 393–396, 2003.

Kimula, M.: *How to produce subharmonics on the violin.* In: New Music Research 28, 178–184, 1999.

Knothe, K. & Wessels, H.: *Finite Elemente. Eine Einführung für Ingenieure.* 3. Auflage, Springer, 1999.

Krassnitzer, G.: *Multiphonics für Klarinette mit deutschem System und andere zeitgenössische Spielarten.* edition ebenos Verlag Aachen, 2002.

Krieg, R.D. and Key, S.W.: *Transient shell Response by Numerical Time Integration.* In: International Journal for Numerical Methods in Engineering, 7, 273–286, 1973.

Knott, G.A.: *A modal analysis of the violin using MAC/NASTRAN and PATRAN.* MSc thesis, Monterey, CA, 1987.

Lai, J.C.S. & Burgess, M.A.: *Radiation efficiency of acoustic guitars.* In: Journal of the Acoustical Society of America 88, 1222–7, 1990.

Le Pichon, A., Berge, S. & Chaigne, A.: *Comparison between Experimental and Predicted Radiation of a Guitar.* Acustica – acta acustica 84, 136–145, 1998.

Lee, M.-H., Lee, J.-N. & Soh, K.-S.: *Chaos in Segments from Korean Traditional Singing and Western Singing.* In: Journal of the Acoustical Society of America 103, 1175–82, 1998.

Leissa, U.: *Vibration of plates.* Publication of the Acoustical Society of America, NY 1993.

Leissa, U.: *Vibration of shells.* Publication of the Acoustical Society of America, NY 1993.

Letowski, T. & Bartz, J.: *Obiektywne Krytzeria Jokosci Pudel Gitarowych.* In: Arch. Akust. 6,1, 37, 1971.

Lipps, Th.: *Grundlegung der Asthetik. Erster Teil.* Verlag von Leopold Voss, Hamburg und Leipzig, 1903.

Luce, D. & Clark, M.: *Durations of attack transients of nonpercussive orchestral instruments.* In: Journal of the Audio Engineering Society 13, 194–199, 1965.

Meyer, J.: *Verbesserung der Klangqualität von Gitarren aufgrund systematischer Untersuchungen ihres Schwingungsverhaltens. (Improvement of the sound quality of guitars due to systematic investigations of their vibrating behaviours)* PTB-Report, Forschungsvorhaben Nr. 4490, 1980.

McIntyre, M.E. & Woodhouse, J.: *The Influence of Geometry on linear Damping.* In: Acustica 39, 209–224, 1978.

McIntyre, M.E. & Woodhouse, J.: *Fundamentals of bowed-string dynamics.* In: Acustica 43, 93–108, 1979.

McIntyre, M.E., Schumacher, R.T. & Woodhouse, J.: *Aperiodicity in bowed-string modtion.* In: Acustica 49, 13–32, 1981.

Meisner, C.W., Thorne, K.S. & Wheeler, J.A.: *Gravitation.* San Freeman, Francisco, NY, 3. Auflage 1973.

Molin, N.E., Wåhlin, A.O. & Jansson, E.V.: *Transient wave response of the voilin body* In: Journal of the Acoustical Society of America 88, 2479–2481, 1990.

Molin, N.E., Wåhlin, A.O. & Jansson, E.V.: *Transient wave response of the voilin body revisited.* In: Journal of the Acoustical Society of America 90, 4, 2192–2195, 1991.

Moon, F.C.: *Chaotic and Fractal dynamics.* NY 1992.

Morse, Ph. & Ingard, U.: *Theoretical Acoustics.* Princeton Univ. Press, 1983.

Neubauer, J., Edgerton, M. & Herzel, H: *Nonlinear Phenomena in Contemporary Vocal Music* In: Journal of Voice, 18,1, S. 1–12 2004.

Newland, D.E.: *Mechanical vibration analysis and computation.* Longman Scientific, NY, 1989.

Newmark, N.M.: *A method of computation for structural dynamics.* In: Proceedings of ASCE, Journal of Engineering Mechanics, 85, 67-94, 1959.

Oertel jr, H.: *Prandtl – Führer durch die Strömungslehre. Grundlagen und Phänomene.* Vieweg, 11. Auflage, 2002.

Olivier, S. & Dalmont J.-P.: *Experimental investiagion of clarinet reed operation and its consequence on the non-linear characteristics of the mouthpiece.* In: Proceedings of the Stockholm Musical Acoustics Conference 2003, 283–286, 2003.

Pakarinen, J., Karjalainen, M. & Valimaki, V.: *Modeling and real-time synthesis of the kantele using distributed tension modulation.* In: Proceedings of the Stockholm Musical Acoustics Conference 2003, 409–12, 2003.

Pickering, N.C.: *Physical properties of violin strings.* In: Journal of the Catgut Acoustical Society 44, 6–8, 1985.

Pickering, N.C.: *Noninear behavior in overwound violin strings.* In: Journal of the Catugut Acoustical Society, 1, 3 46–50, 1989.

Pickering, N.C.: *The bowed string.* Ameron, Mattituck, NY, 1992.

Pitteroff, R. & Woodhouse, J.: *Mechanics of the Contact Area Between a Violin Bow and a String. Part I: Reflection and Transmission Behaviour.* In: Acustica – acta acustica, 84, 543–562, 1998.

Pitteroff, R. & Woodhouse, J.: *Mechanics of the Contact Area Between a Violin Bow and a String. Part II: Simulating the Bowed String.* In: Acustica – acta acustica, 84, 543–562, 1998.

Pitteroff, R. & Woodhouse, J.: *Mechanics of the Contact Area Between a Violin Bow and a String. Part III: Parameter Dependence.* In: Acustica – acta acustica, 84, 543–562, 1998.

Rayleigh, J.W.S.: *The Theory of Sound.* Volume One, Dover Reprint of 1877, 1945.

Reuter, Ch.: *Der Einschwingvorgang nichtperkussiver Musikinstrumente.* Frankfurt a.M. 1995.

Richardson, B.E. & Roberts, G.W.: *The adjustment of mode frequencies in guitars: A study by means of holygraphic interferometry and finite element analysis.* In: Proceedings of Stockholm Musical Acoustics Conference 1983. 285–302, 1985.

Richardson, B.E. & Roberts, G.W.: *The adjustment of mode frequencies in guitars: a study by means of holographic interferometry and finite element analysis.* In: Proceedings of the Stockhom Musical Acoustics Conference 1983, 285–302, 1985.

Richardson, B.E., Walker, G.P. & Brooke, M.: *Synthesis of guitar tones from fundamental prameters relating to construction.* In: Proceedings of the Institute of Acoustics, 12(1), 757–764, 1990.

Richardson, B.E.: *Guitar models for makers*. In: Proceedings of Stockholm Music Acoustics Conference 2003, S.117–120, 2003.

Roberts, G.W.: *Finite element analysis of the violin, extract from Vibrations of shells and their relevance to musical instruments*. PhD V University College, Cardiff, Wales, UK, 1986.

Rodet, X. & Vergez, C.: *Nonlinear Dynamics in Physical Models: Simple Feedback-Loop Systems and Properties*. In: Computer Music Journal, 23,3, 18–35, 1999.

Rodgers, E.O.: *The Efferct of the Elements of Wood Stiffness on Violin Plate Vibration*. In: Journal of the Acoustical Society of America, 1, 1, Series II, 1988.

Rodgers, O.E.: *Relative Influence of Plate Arching and Plate Thickness Patterns on Violin Back Free Plate Tuning*. In: Journal of the Acoustical Society of America, 1, 6, Serie II, 1990.

Rodgers, O.E.: *Effect on plate frequencies of local wood removal from violin plates suported at the edges*. In: Journal of the Catgut Acoustical Society 2,1(8), 7–11, 1991.

Rossing, T.D.: *Plate vibrations and applications to guitars*. In: Journal of Guitar Acoustics, 6, 65–73, 1984.

Rossing, T.D. & Fletcher, N.H.: *Nonlinear vibrations in plates and gongs*. In: Journal of the Acoustical Society of America 73, 345,351, 1983.

Rossing, T.D., Popp, J. & Polstein, D.:*Acoustical response of guitars*. In: Proceedings of the Stockholm Musical Acoustics Conference 1983. Royal Swedish Academy of Music Stockholm, 311–332, 1985.

Schelleng, J.C.: *The bowed string and the player*. In: Journal of the Acoustical Society of America 53, 26–41, 1973.

Schneider, A.: *Tonhöhe, Skala, Klang*. Bonn 1997.

Schneider, A.: *"Verschmelzung", Tonal Fusion, and Consonance: Carl Stumpf Revisited*. In: Music, Gestalt, Computing. M. Leman (ed.). Springer 1997.

Schumacher, R.T. & Woodhouse, J.: *The transient behaviour of models of bowed-string motion*. In: Chaos 5, 509–523, 1995.

Schumacher, R.T. & Woodhouse, J.: *Computer modelling of violin playing*. In: Contemporary Physics 36, 79–92, 1995.

Skudrzyk, E.: *Die Grundlagen der Akustik*. Springer Verlag, Wien 1954.

Stelzmann, U., Groth, C. & Müller, G.: *FEM für Praktiker – Band 2: Strukturdynamik*. expert-verlag, 2. Auflage, 2001.

Stetson, K.A.: *On Modal coupling in string instrument bodies*. In: Journal of Guitar Aocustics, 3, 23–31, 1981.

Stumpf, C.: *Tonspychologie. Teil 1*. Nachdruck 1965, 1883.

Szwerc, R.P.: *Power flow in coupled bending and longitudinal waves in beams*. In: JASA 107, 6 2000, p. 3186–95.

Tietze & Schenk: Halbleiter-Schaltungselektronik. Berlin, NY, Springer, 2000.

Tinnsten, M. & Carlsson, P.: *Numerical Optimization of Violin Top Plates*. In: Acta Acustica United with Acustica, 88, 278–285, 2002.

Tolonen, T., Välimäki, V & Karjalainen, M.: *Evaluation of Modern Sound Synthesis Methods*. Report 48, Helsinki University of Technology. 1998.

Touzé, C. & Chaigne, A.: *Lyapunov Exponents from Experimental Time Series: Application to Cymbal Vibrations*. In: Acustica, 86, 557–567, 2000.

Tsai, C.G.: *Relating the harmonic-rich sound of the Chinese flute (dizi) to the cubic non-linearity of its membrane*. In: Proceedings of the Stockholm Musical Acoustics Conference 2003, 303-306, 2003.

Urbakh, M., Klafter, J., Gourdon, D. & Israelachvili, J.: *The nonlinear nature of friction*. In: Nature, Vol 430, 525–28.

Vössing, H. & Kummer, J.: *Beobachtung von Periodenverdopplung und Chaos bei der Trompete*. In: Fortschritte der Akustik 916-9, 1993.

Wagner, K.W.: *Einführung in die Lehre von den Schwingungen und Wellen*. Wiesbaden, 1947.

Wilden, I., Herzel, H. Peters, G. & Ternbrock G.: *Subharmonics, Biphonation and Deterministic Chaos in Mammal Vocalization* In: Bioacoustics 9, 171–96, 1998.

Wolfram, S.: *A new kind of Science*. Wolfram Media Inc. 2002.

Woodhouse, J. & Loach, A.R.: *The torsional behaviour of cello strings*. In: Acustica- acta acustica 85, 734–740, 1999.

Woodhouse, J., Schumacher, R.T. & Garoff, S.: *Reconstruction of bowing point friction force in a bowed string*. In: Journal of the Acoustica Socitey of America 108, 357–368, 2000.

Woodhouse, J.: *Bowed String Simulation Using a Thermal Friction Model*. In: Acta Acustica United with Acustica 89, 355–368, 2003.

Woodhouse, J.: *The Transient Behaviour of Guitar Strings*. In: Proceedings of Stockholm Musical Acoustics Conference 2003, 137–140, 2003.

Woodhouse, J.: *Plucked Guitar Transients: Comparison of Measurements and Synthesis*. In: Acustica 90/5, 945,965, 2004.

Wriggers, P.: *Nichtlineare Finite-Element-Methoden*. Springer 2001.

Wriggers, P.: *Computational Contact Mechanics*. John Wiley & Sons, 2002.

Zimmermann, K.: *Nichtlineare Theorien von Membranen und deren Berechnung nach einer Finite-Differenzen-Energie-Methode*. Sonderforschungsbericht 64, Weitgespannte Fläch entragwerke, Mitteilung 76/1985, Institut für Mechanik (Bauwesen), Uni Stuttgart, 1985.

Audio Files and Videos on the Web

Audio Files on the Web
http://www.suul.org/Guitar.html

Tracks

1   Villa Lobos | Praelude 1 | beginning | *steel strings*
2   Timbre of the guitars **top plate** from TRACK 1
3   Timbre of the guitars **back plate** from TRACK 1
4   Timbre of the guitars **ribs** from TRACK 1
5   Timbre of the guitars **neck** from TRACK 1
6   Timbre of the guitars **inclosed air** from TRACK 1
7   Villa Lobos | Praelude 1 | beginning | *neylon strings*
8   Timbre of the guitar **top plate** from TRACK 7
9   Timbre of the guitar **back plate** from TRACK 7
10   Timbre of the guitar **ribs** from TRACK 7
11   Timbre of the guitar **neck** from TRACK 7
12   Timbre of the guitar **inclosed** air from TRACK 7
13   plate $40 \times 40 \times .3$ cm | dampings: $R_v = 10\ e^9$ $R_K = 0.999995$
14   plate $40 \times 40 \times .3$ cm | dampings: $R_v = 10\ e^8$ $R_K = 0.999995$
15   plate $40 \times 40 \times .3$ cm | dampings: $R_v = 10\ e^7$ $R_K = 0.999995$
16   plate $40 \times 40 \times .3$ cm | dampings: $R_v = 10\ e^6$ $R_K = 0.999995$
17   plate $40 \times 40 \times .3$ cm | dampings: $R_v = 10\ e^5$ $R_K = 0.99995$
18   plate $40 \times 40 \times .3$ cm | dampings: $R_v = 10\ e^5$ $R_K = 0.9995$
19   plate $40 \times 40 \times .3$ cm | dampings: $R_v = 10\ e^4$ $R_K = 0.9995$

Videos on the Web
http://www.suul.org/Guitar.html

Tracks

1. Complete view of the guitar which is excited by the string displacements shown in TRACK 2. Side view, sub-view, interior view.
2. The low and the high e string of the guitar are displaced in a trapezoid shape at the same point along the string. The time development shows the waveforms of the two strings. The low E string (73.3 Hz) needs four times longer for a complete cycle than the high e string (293.3 Hz) due to the frequency relationship of $293.3 : 73.3 = 4 : 1$.
3. Representation of the vibration of the guitar top plate which starts with large amplitudes at the beginning of the string oscillation right away.
4. Representation of the vibration of the back plate. He starts with the high-frequency motions transferred by the ribs, and just then after a short time is excited by the low-frequency vibrations of the inclosed air to move with larger amplitudes in a low frequency regime.
5. Representation of the oscillations of the ribs. Like the back plate, it starts moving with the higher frequencies transferred by the top and back plate which here are longitudinal waves. Only after a certain time it is excited by the inclosed air to larger bending movements.
6. Representation of the inclosed air of the guitar. It is shown in three planes here. The topmost plane which is shown at the beginning is the airspace just beneath and parallel to the top plate, the undermost plane is the airspace right over and parallel to the back plate, the mean plane is the airspace in half the ribs height and again parallel to the top plate. The inclosed air reacts in low-frequencies and squeezes the ribs only after some time at its sides, the back plate is excited by it earlier.
7. Representation of the vibrations of the neck. His motions are determined mostly by the oscillations transferred from the top plate, wich are mainly quite high in frequency but also can have low frequency components (compare to the quite even oscillation pattern of the neck in comparison with the radiated sound, Audio-CD TRACK 5 and TRACK 11).
8. Example of an FDM of a plate from the plate part of the MTMS software. The edges are damped here. First, impulses are applied to the plate, then it is excited with several sinusodial waves.

We apologize for the fluctuating sharpness of the videos.

Printing: Krips bv, Meppel
Binding: Stürtz, Würzburg